I Look Well, Don't I?

Roy Cleeter

To Juan

Thanks for your support

Roy

 New Generation Publishing

To my partner Mid

This book is for you

Without your love and support. I would not have survived.

Without your encouragement, faith, and tenacity.

This book would never have been finished.

Acknowledgement

The hardships, suffering and uncertainty for the future being endured by the people of Venezuela; Is only one of the many tragedy's being played out in the world in our present times. Due to the possibility of repercussions against those who helped me overcome, compared to the ones they face, trivial problems. I have altered Names, Dates, and Locations.

Without the intervention and help of these people my story could have ended differently.

I will always remember them with appreciation and gratitude.

Contents

Chapter 1

Bon voyage

Day 1

Waking before sunrise; my whole being tingles with excitement. Within the next hour I will set off, on what should be the greatest experience of my life. To sail across the Atlantic. My partner Mid and I brought on speculation a forty-foot sailing yacht (Mae Lee) anchored in Prickly Bay Grenada, designed by Robert Tucker on eBay in December. We flew out to Grenada in January of the new year. It takes hard work and a large part of our budget to get the yacht fit for the purpose.

Mid worked hard on the boat and is as excited as me about the voyage, but she will not sail Mae back with me. Thinking it a daunting undertaking. She flew back to the UK on the tenth of March. Mid will be with me in spirit. Not wanting me to sail solo I agreed to have a crew. My son Del and Mids Son-in-Law Lloyd volunteered, but it did not work out. Mid relented and gave me her blessing to do the trip solo even though it worried her. With my confidence high and my arrogance knowing no bounds. "What can go wrong?"

At six-thirty a.m. I start Mae's engine. Captain Collins who has become a friend during my time in Grenada, slips her moorings. I had met an experienced sailor called Max, in the Tikki Bar one lunchtime. He had crossed the Atlantic solo three times and was preparing to do it again. He taught me a lot about boats. Gave me advice and fixed Mae's mast. He left a few days ago in his boat Sarah, to see friends. We plan to meet in Tyrell bay, Carriacou, and leave together on Saturday the sixteenth for the UK.

As I back Mae away from the pontoon, then negotiate my way through the yachts at anchor. Collins calls out. "Bon voyage my friend, safe journey." On my way to the channel that will take me out of the bay, he gives me a final wave then leaves. He wishes he could come with me, but his commitments will not allow it. I have met good people here, some I now consider friends. My emotions are mixed to be leaving, I will miss them and Grenada. Everything here suits me.

Before I leave the protection of the bay, I hoist the mainsail. Increasing the revs of the engine brings Mae's speed up to six knots. With the cooling system overhauled and the gearbox and water pump replaced, I need to test the engine and gearbox under load. If they give me any problems, I want to know sooner than later. My plan is to motor sail to Ronde island, over twenty nautical miles from my current position. Everything being well I will anchor at Ronde. Then head for Carriacou in the morning. Spend the night there then set off for the UK. Once I leave Carriacou's shores, it will be a long time before I set foot on dry land again.

The sun is shining, just a few clouds in the sky, a light wind and the sea state is slight. The engine is running well and Grenada's sliding by on Mae's starboard side. Knowing I would not want to leave the helm at this early stage until I am sure of how Mae will perform. I had made coffee in a flask. Now having cleared the bay I pour myself a cup. Until now I had only moved her from Prickly bay to Secret harbour which had been too short a trip to learn if she had any quirks. *(Most boats do)*.

As time drifts by, the initial excitement of setting sail wears off. The steady throb of the engine and the rhythmic motion of the boat has a calming influence on me. Sipping the coffee, I glance at the blue foaming water swirling past Mae's starboard side. Dark grey shapes beneath the surface keep pace with us. A pod of five dolphins are cruising alongside. After a few minutes, they dart away. I thought they had gone, but moments later they come back. Their number increased to eight. Switching on the auto pilot, I grab my tablet and scramble up to the foredeck to film them. Getting there as they take turns to ride the bow wave. I have seen dolphins doing this on wildlife programs *(Who hasn't?)* seeing it in real life does not compare. They stay with me for a while then speed away. Back in the cockpit I feel privileged and elated by the experience.

Mae held her course while I was out on deck, which reassures me the auto pilot works. "Why own a dog and bark yourself?" I Leave it on to do its job. Pouring another coffee, and rolling a cigarette, I sit at the back of the cockpit to enjoy the view. As I approach the northern end of Grenada, I should be able to see Ronde island. Last night I had checked the weather forecast for the next couple of days. It will deteriorate before sunset but improve again before dawn, I expect to be in the bay at Ronde before then. Better to be in a sheltered bay for the night and let it pass. While Mae cruises on I daydream about when I will sail into Brighton marine UK, my home port. My partner Mid, family and friends waiting to greet me as I dock.

This happy state of mind ceases when the engine stops. No coughing or spluttering, no weird noise warning that something is wrong. Mae's speed drops to a crawl. Most of her forward momentum was being provided by the engine, without it she wallows. I lift the sole plate above the engine and look it over hoping that whatever is wrong will be visible and simple to fix. The gauges, oil pressure, water temperature, battery levels, show everything as normal. The water level in the heat exchanger is low, but not enough to have caused the engine to stop. Then I notice a hose on the bottom of the water pump has come adrift. I will have to get down into the engine compartment to re-attach it. The autopilot is an energy monster and will soon drain the batteries. So, I turn it off along with everything else except the radio and Gps.

Convinced I have found the problem, I squeeze into the compartment and leave Mae to her own devices. Working on the engine at sea is a bloody nightmare, it was difficult enough when Mae was in dry dock due to the confined space. Now I must contend with the constant motion, a wave out of sync and with more power slams into the hull, causing me to collide with one of the many sharp bits and pieces sticking out from the engine bloc. They could damage me, and they do, many times. After trying a second time to reconnect the hose and failing. My hands and arms covered in oil, grease, and blood, I haul myself back up into the cockpit for a breather and to bandage my wounds. Checking on Mae's course, she is drifting northwest further from the coastline, we need to go northeast. "Once I fix the engine, I will soon have her back on the right heading."

As I sit binding the cuts on my fingers with gaffa tape. (I tried ordinary plasters they were useless). I notice a fishing boat coming towards me. When it is in hailing distance the skipper calls out. "Have you a problem? Do you want a tow back into St Georges?" Shouting out, I am ok, I wave him off. He shrugs and turns his boat away. The crew give me a friendly wave from the stern then go about their business. As I watch them go, I can't help thinking I might have made a mistake. As I squirm my way back down into the engine compartment, I dismiss the thought.

It takes half an hour of sweating, cursing, and getting banged about before I reconnect the hose. Back in the cockpit I turn the ignition key. The engine spins over but won't fire. After trying a few more times I give up, the batteries are getting weak. If I must spend the night out at sea, using up the remaining power will mean not being able to run the radio, Gps, and navigation lights. Feeling anger and frustration at this turn of events, I need to calm down. Pouring out the last of the coffee from the flask I sit and think about my options. The wind speed is increasing, so is wave height. Not by much but looking over at the dark moody clouds closing in from the east suggests this will change. With only hours of daylight left and without the power of the engine I do not think getting into a safe harbour or bay is going to happen. I am being pushed by the wind and current further out into the Caribbean Sea.

I decide to turn Mae around and head back southeast *(as best I can)* towards Prickly bay. Turning around is a slow frustrating exercise. Mae is a stubborn bitch who takes a lot of cajoling to go anywhere near the course I want her on.

Logic says, it's because I do not know how to handle her yet. I can't help thinking the boat just doesn't want to go back to Prickly Bay. Her previous owner had parked her there, leaving her neglected and unloved for several years. She now has a chance of freedom and is taking it. I get Mae heading a few degrees to the southeast. More towards the east would be better but she will not sail any closer to the wind. As a precaution I reef the mainsail, the weather is deteriorating, along with my confidence. When darkness falls, there is no choice but to sit out at sea and wait until daybreak

Needing to keep Grenada in sight, I "heave to." The theory being Mae will not drift too far from my present position. Never having cause to try this before I am not sure if I have gone about it the right way.

I watch as black clouds snuff out the light from the moon and the stars as they advanced towards me. Covering the sky like a thick dark blanket. Not the darkness you experience ashore in a city or town, this is an impenetrable inky black that wraps itself around the boat. The only relief is the red glow of the cockpit nightlight and an occasional glimpse of the navigation light on the bow. The twinkling lights on Grenada's coastline I used as a point of reference, long since disappeared. Never have I felt so alone and vulnerable.

The wind makes the stays to the mast whine as it passes through them. Waves six to eight feet in height hammer the port bow, then make a sizzling noise like hot fat as they break and run along the side of the hull. The pitch and roll of the boat make me feel queasy. When I check the wind speed, it is twenty-five knots. On the Beaufort scale, this qualifies as a force six, *(wind= a strong breeze, sea state = rough)*. At nine pm I put up the cockpit canopy, leaving the back open. The build-up of heat inside the enclosed space gets stifling. But it will stop the waves that are rushing down the port side from invading the boat. It also gives me a sense of security, false as it is.

Going below I intend to make tea and get something to

eat. As soon as I enter the cabin, I want to throw up, I get out quick. Once back outside it passes. Feeling miserable I huddle down between the portside bulkhead and gas bottle holder, wedging myself in. Thoughts of food and a hot drink forgotten as fear pricks my brain. Mae keeps heeling over to starboard, then with a jolt comes back upright. She goes over so far, I get paranoid about capsizing. Telling myself it's my imagination, I cast the thought from my mind. It will take bigger waves than this to capsize her. Wind speed keeps increasing, and Mae's pitching and rolling gets worse.

It is difficult to stay wedged in my corner. Spray from the odd wave that breaks over the bow splatters over the windscreen and canopy sounding like shrapnel. Checking on the wind speed again, it's now thirty-five knots a force seven *(near gale)*. Wave height thirteen feet and rising. I experienced these conditions twice before, both times in the English Channel. In a boat I know and with a crew. On this boat, an unknown quantity and alone, at night in these conditions. A new experience I could do without, I feel sure something will give. "It does, me!"

Retching I throw up over myself and the cockpit deck. As I try to get my head over the side Mae lurches and I slip on the metal sole plate, banging my head on the ships wheel and falling into the vomit that covers the deck. Mae rolls from side to side which causes me to do the same. Coating me in the foul mess as I roll with the boat and carry on retching. When I have nothing left to give, I lay on my back staring up at the cockpit roof. Exhausted, I want to die, I should keep watch, but I cannot move.

The navigation lights are on and the radar reflector up. So, I should be visible to any other vessel that might come along, not that I care. I focus on the safety hook fixed to the roof. It's making a clicking sound as it swings from side to side. I must be delirious from the bang on the head. It sounds as though Mae is laughing at me. My last conscious thought as sleep claims me is. "You did that on purpose, you bitch".

Day 2

When I wake up, I am still laying on the cockpit floor. It's very rare for me to surface alert and aware of my surroundings. Like a diver coming up from the depths. I take short breaks. Too fast invites disaster. You would think this was an occasion my mind would quickly clear. But no, it sticks to its routine.

Opening my eyes, I see the hook in the roof is still and silent. Mae is no longer rolling from side to side, just rising and falling with a steady rhythm. As I push myself up, I put my hand into a pool of congealed vomit, a rank odour fills my nostrils. The heat generated inside the enclosed space of the cockpit has cooked it into a sour porridge. Everywhere I look, small steaming piles of vomit are fermenting. Unzipping the canopy to let in fresh air I scan the horizon. My heart sinks, Grenada is not in sight. There is nothing just an empty sea. On the good side the weather has improved, and I do not feel sick anymore. I am in a mess like the cockpit and smell as bad. I need to clean up, I can't think of anything else until then.

When I finish, I make tea and roll a cigarette. I cannot check my position. Leaving the Gps on through the night combined with the navigation lights and radio, has drained the batteries. Everything is dead. I will have to wait until the solar panels and wind generator charge them. I hope they can put enough power into them to run everything.

Savouring the tea and watching the sun's rays dance across the sea makes last night seem like a bad dream. I try to convince myself that my fear and imagination had blown the experience out of proportion. "It couldn't have been that bad, could it?" My natural optimism comes into play and I tell myself, "The boat is afloat, you're alive, everything is fine.

I wait until ten a.m. before there is enough power to turn on the Gps. The heads-up display, (HUD), shows me that while I slept Mae turned away from the wind at a right angle and travelled along at an average speed of three knots. I am forty-eight miles southwest from Grenada and eighty odd miles from Trinidad. I need to figure out what to do. Thinking it over I decide to turn around and have another go at getting back to Grenada. It will be difficult, but Grenada is closer than Trinidad. More experienced sailors would sit down at the navigation table and work out all the intricate "This and That's." But I think if I must leave Mae till next year it will make me happier leaving her in Grenada rather than a place I know nothing about. I struggle with Mae to get her turned around and put her on a heading that should take me in Grenada's general direction.

Unfurling the headsail and taking the reef off the main. Mae's speed increases to five knots. As I am making my approach from a different angle, I hope the wind and the current will not be such a problem. I know I will run parallel to the island when I draw level with it, but I think I should not be too far out from the coast. With luck, I could meet another boat and get help. Satisfied Mae will hold her course I make breakfast and coffee then come back up to the cockpit.

I must stop myself checking the Gps for course and distance to Grenada. Watching the heads-up display show the range decrease and the off-course track increase depresses me. Mae is heading too far towards northwest, I need her on a more northeast track. No matter how much I tweak and fuss she will not sail closer to the wind. Midday the radio bursts into life. "Mae Lee, Mae Lee, Mae Lee, this is Oscar, Oscar, Oscar are you receiving over."

Rushing down into the cabin I grab the mic and answer. Still a long way from Grenada I am amazed to be receiving him. Hearing no radio traffic from other vessels since setting off, I was thinking the bloody thing had packed up. He asked me where I am and if I'm ok. I assure him I am, leaving out the part about throwing up. It's too embarrassing to mention. I give him my position and tell him what happened with the engine. "You only have sail, no engine?" "You are sure you cannot get it going again." I confirm this. "Ok," he says. "You will have a difficult time." "The wind and the current are against you and the weather forecast is not good." Telling him, it had been rough last night. He says. "Tonight, expect more of the same." "Do the best you can to get back, and I will look at the engine for you." I resisted the urge to be sarcastic about, "The difficult time."

Making my way back to the cockpit I have mixed emotions. Hearing Oscar's voice and able to tell him my situation lifted my spirits. But the thought of another night at sea didn't make me jump for joy. Knowing I've got no chance of being in a nice safe anchorage before the end of the day. At first, I console myself with the thought it could be worse. At least I'm not in the Atlantic in huge seas being tossed around. A voice in my head says. "Everything happens for a reason." "But then maybe not?" There are too many variables, and I am getting a headache thinking about it. The one certainty is my adventures over before it has begun. No way can I get back to Grenada, have the engine fixed and leave before the hurricane season.

I need to prepare Mae for when the weather turns. Hearing the sails flapping about. I look up at the main, it is ok. The headsail is making all the noise. The sheets, *(ropes)* to it somehow have got tangled with each other preventing the sail from extending out and being filled by the wind.

The weather has improved, but it's still a bumpy ride. The winds still strong, but better than the previous night. A heavy gust sweeps across the deck forcing the sail to wrap itself around the furler. The ropes rear up and writhe around. No way do I fancy going out on deck, but I have no choice, who else will sort it out?

Lashing the steering to keep Mae on course I pray she will remain stable, not lurching around as she had last night. After putting on the safety harness I climb up onto the foredeck. I cannot clip myself onto the safety wire that runs the length of the deck until I negotiate my way over the dinghy's secured midway. Hanging on to the ropes that lash them down and keeping low I inch my way along. To fall overboard doesn't bear thinking about. I have done nothing like this before with no one to give me a hand or advice, no one I can rely on in case of a problem. This situation has put a spotlight on my failure to recognise my inexperience and appreciation of sailing solo. I stop guessing about the what ifs, buts and maybe's. That will get me nowhere

As I get closer to the bow, the wind gusts catching the ropes and whips them about in a frenzy. Staying low to the deck trying to keep out of their way; I wait for a lull in the wind before standing up and making a grab for them. When my chance comes, I reach out to the ropes. Another gust snatches them away and causes the sail to swing from side to side. The ropes coil up like two snakes about to strike, and strike they do. One hits me in the head and the other lashes my back like a whip. The sail billows out then swings back, pushing against me. Without my harness secured, I would be thrown over the side. Once I get control of the ropes and untangle them, I keep tension on them as I crawl back to the cockpit. With each gust of wind, the sail tugs on the ropes with such force my hands sear with pain as it tries to rip them from my grasp.

Dropping back into the cockpit, I reset the headsail and secure the ropes. While I was up on deck Mae as usual has gone her own way. Putting her back on course, *(well as far as she would let me)* I then sit down and treat my wounds. Raw and blistered rope burns cross the palms of my hands, my head aches, and my back is sore from the thrashing I received. Unbelievably I feel good about myself, I faced a problem. "A small one admittedly," but I dealt with it. Trying not to think about how far out from Grenada's coast I will be when I draw level, not wanting any dark thoughts spoiling my feeling of wellbeing, I keep busy cleaning the main cabin and cockpit *(they are still a mess from the night before)*.

Just before dusk I make my evening meal. As I watch the sun dip below the horizon, I feel the wind increase enough to make me look to the east. Dark clouds are forming in the distance and heading towards me. The bad weather Oscar warned me about is on its way. After a few hours the clouds are above me, blanking out the stars. All around me is black, but for the occasional flash of white foam as a wave breaks against the boat. With the darkness comes the wind, like a tube train approaching a station. A low distant rumble getting louder, and then a roar as it bursts from the mouth of the tunnel.

The wind increases so does wave height. Thinking it prudent I make a flask of coffee. Mae is already rocking, and I know it will get worse. Trying to make a hot drink would be a nightmare.

The memory of last night in the cabin is fresh in my mind. I don't want to spill my guts out again and suffer cleaning it up. Everything except the navigation lights and radio, I turn off. Then set the alarm on my mobile to go off every hour in case I fall asleep. I need to keep a check on Mae's heading and look around for other vessels.

The wind and wave action intensifies and soon I am in the top end of a force seven *(near gale)* for the second night running. I cannot see a damn thing. The howl of the wind and the fizzing sound of big waves sweeping along the hull and breaking astern sends fear crawling over me. Mae heels over at an alarming angle the whole thing freaking me out. Once again, I wedge myself in by the gas bottle and hold on. It takes a lot of effort to stop myself being thrown across the cockpit.

A voice on the radio startles me. Whoever it is speaks Spanish. The only thing I understand, is his vessel is in trouble. About an hour later he speaks again, this time in English. His voice has an edge to it. "Any boat." "Any boat." "This is," *(I can't make out the name due to static)*. "We have no power." "Our engines are down." I try to contact him, but my calls are not acknowledged. There is nothing I can do. I hear no more, I feel guilty but console myself that whoever had been calling was not alone. From the bits of information that got through the static, I think he is on a freighter. Not that it makes much difference. If you are in trouble and your boat is at the mercy of a hostile sea, under threat of sinking, it doesn't matter how big it is or who you're with. If no one can help, you are on your own.

Day 3

As the sun comes up the weather is frisky. Mae is still bouncing around, the ride uncomfortable. Turning off the navigation lights I check the batteries. As I expected they are low on power. So, I cannot turn on the Gps to check my position. I make breakfast. Bran Flakes, half Grapefruit, and a cup of tea. Mae stocked with enough food and water for three people to cross the Atlantic to the Azores, means I will not starve or die of thirst. Tinned soup, meat, and fish. Rice and noodles. An inordinate amount of macaroni, *(God knows why?)* fresh fruit and vegetables. Plenty of tea, coffee, concentrated orange, and powdered milk I have stored in her lockers. Plus, ten litres of UHT milk. Because I cannot run the fridge when I open a carton it only lasts a day, then turns into a foul evil smelling yogurt. There are three hundred and eighteen litres of water in the starboard tank, also seven fourteen litre water jugs and two twenty-three litre Demi-Johns. Like the ones you see in an office.

The portside water tank is empty. It leaks and needs replacing. I busy myself through the morning checking Mae over. Making a list of all the things that need doing every day. Bilges, Rigging, etc.

The bilges only suffer a slight ingress of water. But I still need to pump them out, I don't want it to build up. The navigation lights seem to use an excessive amount of power through the night. I suspect something else is draining the batteries, but I do not have a clue what it can be.

There is enough power to turn on the Gps at midday, disappointed by the read out on the HUD is an understatement. Even though Mae has travelled fifty miles. The wind and the current forced her even further west during the night! We are still forty miles from Grenada

My inexperience in navigation and interpreting the information on the Gps keeps letting me down. I resign myself to the fact I will spend another night at sea. During the day the bad weather eases, but not by much. When darkness falls, it isn't long before the wind meter is showing force seven on its screen. By eight pm the power of the wind and waves increase, and I am wedged back into my usual place. Suffering the emotional rollercoaster, the night hours put me through. Alternating between fear, self-pity and hatred of the engine and my own stupidity that has caused this situation. *(I should have excepted the offer of a tow from the skipper of the fishing boat.)* At nine pm I realise I have left the Gps on.

Standing to turn it off Mae lurches and I am thrown against the handrail next to the helm; it catches me under my rib cage and knocks the wind out of me. The last thing I need is an injury. Now I am not suffering from the effects of seasickness I risk going down into the cabin. Thinking it might be safer there. I will still have to go back up to the cockpit to look around. So, I set the alarm then laid on the bunk.

I am not sure if being in the cabin is worse, hearing the waves rush along the side of the hull, a weird gurgling noise that sounds as if it is inside the cabin with me. Sometimes Mae heels over and there is a loud bang as if someone's outside hitting her with a sledge hammer, trying to force their way in. Mae shudders, comes back upright then pitches forward. Creaking and complaining about the abuse to her hull. Once the action is so bad I get tossed out of the bunk even though I have a lee cloth up to hold me in. It's a sobering thought, between me and the mayhem outside is just six millimetres of steel plate. Less, on some parts of the hull. The weather is punishing me for my naivety and lack of experience. I promise myself, when, if; I make it back to the UK I will sell Mae, buy a cabin cruiser and just potter about inland on rivers and canals. With that thought in my head I fall asleep.

Chapter 2

Tea & Diesel

Day 4

Not hearing the four o'clock alarm, I do not wake up till dawn. Left to ring it killed the phone's battery. The first thing I realize is the boat is not pitching and rolling, just gently rocking. A beautiful calm day greets me. A few clouds, drift across the sky, the wind is light and wave height negligible. The dark angry storm clouds have gone. After making tea and breakfast I take it up to the cockpit; Unzipping the canopy, I sit on the roof of the aft cabin to eat and enjoy the moment. It is short lived. When I take my first sip of tea, I gag; It tastes disgusting.

Back in the cabin I pump the tap. The water comes out a light brown colour. The more I pump the darker it gets. Dipping my finger into the flow it feels oily. Holding my finger to my nose I smell diesel. "Oh, Great!" How the hell had diesel got into the water tank? I went out on deck and checked the tank. The dipstick has a sheen to it. Wonderful! I have been drinking contaminated water since leaving Grenada. Diesel must have been floating on the top. With the rough weather tossing the boat around, gradually it mixed with the water. The contamination must have occurred when I filled the tank. It is not possible for it to happen while at sea. There is no connection between the fuel and water tanks. Why would there be?

What effect drinking diesel for three days will have on me, I have no idea? There is no point in worrying. There's nothing I can do. I laugh. Thinking about the long term seems optimistic. Now I only have water in the jerry jugs and the two containers. This amounts to one hundred and forty litres, the water in the tank is of no use, even to wash with. Unless I fancy a nice dose of dermatitis.

It is not a given I can make it back to Grenada within the next twenty- four hours. Anything could happen. Water has become a precious commodity, I cannot afford to waste any.

Using a spare manual tap and fixing a length of pipe to it, I can siphon water from a jug to the kettle. It saves spilling any. I also fill a one litre water bottle. When I take a sip, it tastes strange. Standing in a plastic container in thirty-six degrees of heat it has a right too. I know in these temperatures I need to drink at least three litres a day. Two litres I flavour with orange squash and make the other up with tea. Boiled for tea its fine but it has an after taste.

This turn of events added to everything else that has happened over the last three days, turns my mood dark. Not to the point of depression, but the reality of my situation is taking large bites out of any optimism I have left. I know I must keep busy. After a strip wash in sea water, I tidy up the boat and go through my daily checklist. By midday the batteries have enough power to turn on the Gps. The HUD reading is not good. Even though I am now level with the northern end of Grenada, I'm fifty miles west of it.

In a mindless rage I stomp around the cockpit throwing the toys out of my pram, cursing the wind and the boat. Jumping up and down on the sole plate covering the engine I scream and shout.

When I finish my tantrum and calm down I know it's me who is to blame, but I had not wanted to admit it. Putting Mae on a starboard tack I get her sailing as close to the wind as she will go. Mae is making four knots, but I soon realise any gain I make towards Grenada is being offset by the current. "I must keep trying", "what else can I do?"

The rest of the day passes by uneventfully and as the sun drops into the sea. Darkness creeps up and wraps itself around the boat. While I sit drinking tea and chain smoking, listening for the distant howl of the wind which will signal another night of anxiety and fear. My mind conjures up different scenarios of how this voyage could end. None of them have a happy conclusion. In my battle with Mae and the elements I have brought a knife to a gunfight, and I suspect there can only be one outcome.

The roaring wind I fear does not arrive. Mother nature

has given me a reprieve. Even Mae who is now on a port tack behaves herself. There's nothing to do but keep watch. The night sky filled with stars; is a magnificent sight now there are no clouds to block them from view. Doing a three-hundred and sixty-degree scan of the horizon I see no other vessels. Just empty glittering sea. Needing to sleep, at nine pm I lash the helm, and go down into the cabin. As I haven't seen another boat for the last three days, I reset the alarm to ring every three hours and lay on my bunk. I'm willing to risk it. Besides, I think I am too far west to come across anyone island hopping. Normally before I fall asleep, my imagination runs riot predicting disasters that wait to befall me. Mercifully not tonight. Closing my eyes, I drift off.

Day 5

A faint beeping penetrates my subconscious. The phone with the last of its power encourages me to crawl out from the safety of sleep into reality. Its six a.m.; I missed the three a.m. alarm call. Mae is still afloat, I'm not dead. So, I lay on the bunk for another five minutes before going up to the cockpit and looking around. When I do, there is still just an empty sea. Mae's speed has improved, but other than that nothing has changed. Back in the cabin I make breakfast. When I finish eating, I move up to the cockpit and smoke a cigarette. My mood is lighter, more relaxed, I cannot explain why this should be. The situation still concerns me, but my feelings have lost their edgy quality.

During the night, the only equipment using power is the navigation lights and radio. The solar panels and wind generator must create enough charge to handle them. But once again on checking the batteries, the power drain seems excessive. It's a pain in the arse having to wait for the power levels to rise, so I can turn on the Gps and check Mae's position. I could get an approximate idea by working it out on the chart, but I'm not confident about my knowledge of the formula. I have taken the theory test for the day captains' ticket and passed. But sitting in a classroom working out position was a breeze; a mistake would not kill you, out here it could. I need to study and learn. If anything happens to the Gps or I have no source of power for it; I am fucked. That thought makes me laugh, "wasn't I already?" After going about the daily routine of checking over the boat, I then sit on top of the aft cabin with a cup of coffee and a cigarette.

An ample supply of tobacco, papers and filters is onboard. Smoking is bad for my health, I know that. But dragging on a cigarette helps me to focus and relax in stressful situations. As crazy as it sounds I would view losing the means to have a smoke as much of a disaster as losing the Gps.

The weather is glorious, and makes me feel good, but there is a trade-off. With the light wind, Mae is slowing down from four knots to just under two. Plodding along, rocking from side to side like an old lady with arthritic hips. As I look ahead a faint outline forms on the horizon. I cannot figure out if it is a low-lying cloud or land. Looking through the binoculars I am still too far away to make out which it is. I will have to wait until Mae gets closer. What or where ever it is, Mae is heading straight for it. As much as I want it to be land there is no chance it's Grenada.

While I sit on top of the aft cabin with my eyes glued to the binoculars, I think about the past few days. Ever since the engine died I have been living in a "Woe is me world." Each new problem couldn't have been that bad, could it? Letting fear get the better of me, my mind must have blown everything out of proportion. This new calm mood helps convince me of it. Time to stop being a wimp! Banish fear and doubt from my thoughts. Imagining my father standing in front of me, I think of what he would say right now. "A miracle will not happen; You will not get back to Grenada; You cannot fight the wind and the current." "The AA is not coming to fix the engine and your mother will not rescue you. Work out where the boat can get to and stop whining." "Now get on with it!"

He had believed in tough love. *(Dying when I was*

twenty-six. He was forty-four). I left home just before my sixteenth birthday. Nine months later pitching up on my parent's doorstep, homeless, broke, and only owning the clothes on my back. It was six pm on a Wednesday night and pouring with rain. Expectation of a warm welcome like "the return of the prodigal son" dissolved when my father opened the door.

He looked me up and down. "Yes, what do you want?" I explained my situation. Instead of letting me in, he gave me a hard stare and said. "Get a bloody haircut, tidy yourself up and get a job. Then come back and I might think about letting you in, goodbye," and shut the door. At that moment, I did not like him very much, but later in life thanked him.

He's method might seem harsh to the modern way of thinking, but I know he had done me a favour. There were a few times in my life when he gave me a mental kick up the arse. But never with malice. He loved me and did it to teach me to be self-reliant and face up to life and whatever it threw at me. Feeling more focused after giving myself a talking to, or maybe the voice in my head had been my father, who knows?

I still did not have enough power to turn on the Gps, but I remembered a device onboard that did not need the power the Gps required. With all that had happened over the last few days I had forgotten about it. It would give me longitude and latitude. How accurate it is, I have no idea? Using it with the chart I should be able to work out my position. Even if it is approximate, it's better than nothing. *"I am relying too much on the Gps anyway."*

I think I am heading towards a group of islands known as "Los Testigos," *(Witness islands in English)*. They come under Venezuelan authority. When I can turn it on, the Gps should confirm it. As Mae limps closer, I make out more detail through the binoculars. The largest island "Testigo Grande" is dead ahead. I can see two others which I assume to be "Conejo" to the east and "Iguana" to the southwest. The chart shows three more islands in the group, but they are not in view.

As much as I want to feel dry land under my feet I feel uneasy. The closer I get the more these feelings increase. The coastline does not look welcoming. There are patches of surf breaking just in front of the island which I take to be reefs. There are no manmade structures in view, just high cliffs, and scrub vegetation. I am still too far away to be sure. To be honest, I don't feel I want to get any closer. These islands seem to scream at me, "Stay Away!"

The only knowledge I have of Venezuela and its territories is from what I learned from a few people who sailed their yachts back from there, and snippets of news reports. Yacht owners who have lived there told me. "Venezuela is a beautiful country, a dirt-cheap place to live, as long as you can put up with the corruption, and violence, *(which is rampant)."* Now they feel it is far too dangerous and not worth the risk. There are news reports that the situation is getting worse due to the failing economy.

There are food shortages and the Americans have an embargo in place. Also, you cannot trust the military and some Venezuelan fishermen are not averse to piracy. This is hearsay, but for a lone sailor who has no contact with the outside world, this group of islands does not seem the place I want to be.

When the Gps powers up. It shows me I am right. Los Testigos are the islands ahead, and I am three miles out from the largest one. The chart loaded into the Gps comes with useful information on each land mass on it. I know it is up to date, I brought the latest one in Grenada.

The islands have a scattered semi-permanent population of two hundred, who are fishermen, also a Venezuelan naval base is on the south side of Testigo Grande. I am approaching the north side which is uninhabited. There is a cautionary note. Confirmed reports of piracy in the area. Some incidents resulted in the death of sailors.

I Consider my options. If attacked, I could not outrun them. My only means of defence is, a diver's knife, two machete's and a flare pistol. Not what you can call an arsenal. They would have at least pistols and or AK47s.

That was enough for me, I do not consider myself to be a coward, but I am not brave either. So, I opt for Plan B. *"(Run away, run away)."* Mae must agree with my decision because she turns without her usual fuss and heads north.

It isn't long before the wind picks up and her speed increases from a dawdle to four knots. For the rest of the day I scan around for any vessels heading towards me. Not until the Gps reading shows I am twenty miles away from Testigo, do I relax. Until then if any boat had appeared I would have viewed it with suspicion. After studying the charts, I decide my best option is to head for St Martin. I cannot afford to pissball around anymore.

The windward islands are on a slight curve, most of them too far east. St Martin is North East. Three hundred and ninety miles away in a straight line. I am under no illusion I will cover a lot more miles than that. I base my calculations on it being a five-hundred-mile trip. If I can keep an average speed of four knots and have no mishaps, I should be there in five to six days.

Confident I have made the right decision, I still as a precaution ration myself with water. Having just over one hundred and thirty-six litres, I will still allow myself three litres a day. That will give me a more than enough safety margin. Just as well I have! The wind soon eases off and Mae's speed drops to two knots. At this speed, it will take me twelve days.

I am keeping a video diary, talking to Mid as if she is with me helps. Alone in the dark and the lack of human contact is getting to me. If I could hear chatter on the radio, it would not be so bad. There has not been a sound from it since the night I heard the guy calling for help. If I had made it into the Atlantic heading back to the UK. There would have been lots of lonely days and nights. That would have been different. At least I would achieve my goal. This is a bloody disaster. The alarm is reset to go off every hour. The luxury of the three-hour break is over. I am worried about encountering a hostile vessel. But when I drag myself off the bunk and look around there is nothing, and the night passes without incident.

Day 6

A short time after sunrise I get visitors, a pod of dolphins followed by a lone Gannet. They do not hang around long but are a pleasant distraction. The wind had picked up overnight and Mae's speed is back to four knots. I go through my routine then make breakfast and sit in the cockpit to eat. Mae occasionally buries her nose into the waves and spray splatters over the foredeck. The wind has shifted, and she is facing more into them. The ride is uncomfortable, but I do not mind I am heading in the general direction I want to go.

Feeling good, I am enjoying myself despite my troubles. The temperature is thirty-six degrees; the breeze does not have much effect. Inside the cabin, it is forty degrees and very humid which is why I eat most meals outside. As sunburn is no longer an issue, I am naked. Not because I am turning into a naturist. It's a relief not to have the sweaty salt encrusted waistband of my shorts chafing my skin. I keep them close to hand to put back on. Just in case I meet another boat. After washing in sea water every morning, I treat my skin with moisturiser. Mid insisted that plenty was on board. During the day, time passes quickly. There is always something to do. I have developed a habit of talking to Mae and the sea regularly. When washing up my plate and cup, I lean over the side of the boat and respectfully ask the sea to send a wave high enough to rinse them, which it always does. I then thank it, I have this mad idea that if I am nice to it, it will be nice to me.

The one-sided conversations with Mae are of a more abrasive nature. I feel we are in a struggle for dominance. She cannot answer back. *(How can she? It's a boat. An inanimate object)* But I am convinced I can feel her disapproval when I speak to her harshly. On the flip-side I am sure she purrs like a cat when I am being nice.

I find it strange that during the day I am wide awake even

though I only have about four hours' broken sleep of a night. Time at sea in hours so far, work out to one-hundred and twelve. Only having slept for eighteen. How long can I keep it up? An Irish sailor I met in the boatyard on Grenada told me that sleep depredation creeps up on you without you realising. hallucinating and weird thoughts are a symptom. He experienced it sailing solo from Cape Verdi to the windward islands. Maybe it is already happening? Of a night, time drags. In the darkness, my mind is assaulted with doom laden thoughts. I cannot help it even with my newfound sense of wellbeing, they try to worm their way into my head. One always finds a way in. The fear of Mae taking a knock down or capsizing during the night. For it to happen during the day would be bad enough. But in the dark. That holds a special terror for me.

Chapter 3

Becalmed with Problems

Day 7

The alarm on the mobile wakes me up. Turning it off a new sound and different aspect of the boat registers in my brain. Instead of the gentle gurgle of water passing along the side of the hull, there is a rushing noise like surf breaking. Mae's heeled over. Not to starboard as she should be, but to port. Realizing what has happened I shout at her. "You fucking bitch, you've turned yourself around". As if in answer for cursing her, a strange creaking sound comes from where the mast sits above the cabin. Looking up at the clock on the bulkhead, I see its five a.m., and go on deck.

Outside, her sails filled with wind, Mae is flying along at a fair lick. Which would be ok if she was going in the right direction? With dawn fast approaching, I turn off the navigation lights and switch on the Gps, hoping there is enough juice in the batteries to power it up. The wind generator blades are spinning so fast, I cannot see why there wouldn't be. The heads-up display flickers and stays on just long enough to show me while I slept we travelled ten miles back towards Los Testigos.

Angry with myself for not waking up and furious with Mae, I call her names which I know is irrational. I must turn her. She responds to this by coming upright then heeling over, catching me off balance. Falling against the grab rail with such force the wind gets knocked out of me. "All right, all right, I am sorry I called you a fucking bitch".

Mae comes upright and tries to throw me again. "Ok, ok. I'm sorry about all the other names I called you as well". She accepts my apology and stops trying to kill me. Coming about she resumes our course north at a speed of five knots. Logic tells me it was my inept actions turning her around which caused the boats response, but in my sleep deprived state I am not so sure?

Once I feel she will stay on course *(within reason)*, I relax and go about my morning routine. Checking the batteries, the bilges, etc. etc. I notice the back stays to the mast are waving around, the tension has gone out of them. Assuming the bottle screws have loosened off, I get the tool needed to tighten them. It is not the screws. They are tight. So why have the stays gone slack?

The rigging got overhauled, and double checked in the boatyard. Sitting on the aft cabin roof I see the problem staring me in the face. The boom is sitting on top of the coach roof. The mast leaning back towards the stern. Putting on a safety harness I make my way to the bow. The sight of the forward chain-plate peeled back from the deck holding on by an inch of steel makes me feel sick. If it shears off the consequences do not bear thinking about. I will lose the mast.

I must drop the sails, the pressure they are putting on that small piece of steel is immense. It will not be long before it splits away from the deck. The swell of the sea is increasing, and Mae is pitching her nose into the waves like a pig hunting for truffles.

My initial horror leaves me. Controlled panic takes over. Telling myself to calm down and "Do what you can." "What will be, will be." I make my way back to the mast and drop the main-sail. Back in the cockpit I release the ropes to the headsail and heave on the furling line to roll it back in.

I must strap the chain plate down. "How!" After making three trips to the bow taking anything I can think of to help achieve this, a three-foot length of 3"x 2" timber, plenty of rope and a strap is piled on deck. The timber and the strap are on the boat because Lloyd used them to lift the engine when he replaced the gearbox.

Loss of forward momentum causes Mae to pitch and roll. This will not be easy. Tying a length of rope to one end of the 4" x 2" I then jam it onto the chainplate. I have to lay on the deck with my head over the bow. Then, I can swing the loose end of the rope underneath it as Mae rides up a wave. I must be quick or else the following wave will catch the rope and drag it down.

Mae sticks her nose in the water taking me down with her. I come back up coughing and spluttering. This happens three more times before I get the timing right. When I do, I pull tight and tie the rope to the other end of the timber.

I must go through the process again with the strap. This time when I reach out to steady myself, my hand catches between the plate and the furler drum as it bears down on the deck. The sharp edges of the plate slices into my fingers and the split pins rip across my hand as I pull it away. The pain is so bad I want to scream, but I keep my mouth shut. Mae has pitched into a wave and my head is under water. My lungs full of Caribbean Sea, is a bonus I do not need.

After catching my breath, I make use of the three spare halyards/ropes attached to the mast. Securing them as far forward as I can they should help prevent the mast from tilting back further. Looking at my effort, it is not pretty, but it's the best I can do. I must sort out my hand. My crushed fingers are throbbing with pain and blood is dripping everywhere. After washing in sea water, I bind them with duct tape. I need tea and a cigarette. Going down into the cabin, another disaster is waiting for me.

The two five-gallon Demi-johns of water stored under the table have tipped over. Most of the contents are sloshing about on the cabin floor and leaking into the bilge. There is just over a gallon left in each.

Too tired to do anything about it I make tea and have a smoke. I become aware the wind has dropped, the sea is like a millpond. My wind meter registers zero, and the temperature goes up to forty degrees. Only Mae's, creaking, and groaning breaks the silence. After a short time, even that stops. I feel like shit! Stress and the effects of dehydration are taking their toll. I do not know what to do. So, I turn on the navigation lights even though it is only dusk and fall asleep

Days 8,9,10,11

I am drifting along a winding pathway feeling content; Enjoying a sense of warmth and security. Sensing the mist before I see it, my mind fills with apprehension. Tendrils of vapour slither over the path, not aimlessly, but with a purpose. They wrap around each other like lovers, advancing toward me. Their lanquid embrace becomes a frenzied, chaotic dance. A swirling, boiling mass. My sense of contentment is banished from my soul and replaced with anguish and dismay.

Mids face appears before me contorted in agony. She is struggling to say something, but when she opens her mouth, the vapour pours in, choking off her voice.

With an effort she shouts. "WAKE UP YOU SILLY BASTARD!"

As soon as I open my eyes a dazzling glaring yellow light, and a wave of intense heat pierces my eyes, threatening to boil them in their sockets. I do not know which is worse, the nightmare I have just left or the reality I now discover myself in. Now I am aware, my brain registers the aches and pains coursing through my body and I remember.

I am on a crippled boat, becalmed and miles from land, adrift in the Caribbean Sea. The sarcastic side of myself gives me a subconscious stream of abuse. "Oh, deep joy, what a laugh, got any brighter ideas?" "I mean, besides the one where you buy a boat and get stuck in the middle of the Caribbean." Groaning with pain, I pull myself up out of the corner of the cockpit where I dozed off. I laid there so long my joints have seized up.

The world spins and blurs before my eyes as I stand up. My head throbs, and I feel sick. Stumbling across the deck to the companionway I have to hold on to the helm for a few minutes until the sensation passes. The voice in my head continues to berate me, babbling away just beneath the surface of my mind. Mentally I shout at it to SHUT UP! Banishing it back to the shadowy little corner of my mind were pessimism and self-recrimination live.

I need to get liquid inside me. Once I feel able to move I get a litre of orange squash, then rest on the bunk and sip it. I resist the impulse to guzzle it down, I know that will do me more harm than good. After a while I feel better and make tea. Going back up to the cockpit I flop into a corner.

It is three thirty in the afternoon. No wonder I feel so rough, I have slept through the hottest part of the day under the full glare of the sun. Over the next three hours I sip my way through two more litres of water and drink more cups of tea. As I smoke a cigarette and watch the sun sink into the sea I shiver. It confirms I am right about having sunstroke. After scanning the horizon, I make my way down into the main cabin wrap a blanket around myself and lay down on a bunk.

Twice during the night, I go up to the cockpit to look around even though it seems pointless, I have given up worrying about anything. Mae is dead in the water and as I have seen no other vessels, I think the chance of having a collision is non-existent. There is not a breath of wind. With only the solar panels charging the batteries I wait until mid-afternoon before I turn on the Gps.

Mae has barely moved. She is just drifting with the current. I am at a loss, what to do? A big horse eye jack cruises around the boat, under Mae's hull there is a small community of fish using her as a shelter, it tries to pick them off.

Dropping a line over the side, within two minutes I have three fish laying on the deck. They look edible. To save gas I cook them on the small barbeque I have on board. This makes a nice change from a diet of tinned soup and noodles.

While eating the fish, I think about my options.

Option one. Sit here and hope another vessel will appear? My water could run out before then and I would be dead from dehydration. Dying of thirst would not be a pleasant death

Option two. Haul the sails back up hope for wind and try to get to St Martin? The mast might collapse. But then again it might not?

If I unfurl the for-sail halfway and haul up the mainsail but only to the third reef. There will be pressure on the mast but not as much as full sail would be. The boom is sitting on top of the coach roof. Will it be a problem when; if, I get underway? No, I can put my shoulder under it when I need to alter the sail.

It did not take long to decide the second option, would be best. At least I would be doing something to help myself.

While waiting for the wind time has no meaning. I catch up on sleep, fish and talk to the tablet. I want a record of events to show to Mid, when, if I survive.

Physically I feel well, my mental state is another matter. Besides talking to the tablet, I hear Mae singing through her stays. Not the high-pitched whine they make in a strong wind. But subtle as if from a distance. Whenever I get close to them, it stops.

Fear and stress do not visit me, in my madness I see myself in a painting hanging on a wall in a gallery while the world passes me by.

A man in a boat on a static sea

Day 12

Early hours of this morning the wind arrives. Sitting here on the aft cabin roof a faint whisper brushes my cheek. When I turn towards it in the distance, I see a ripple on the water. Again, I feel it, stronger this time and just enough to turn the blades of the wind generator and get Mae to move.

Now the generator has kicked in it should not be long before I can turn on the Gps and check my position. My lethargy has gone, to be moving again is great.

It's late in the afternoon and there is enough power for the Gps. The heads-up display shows Mae moving at a speed of one and a half knots and one hundred and seventy-eight miles from St Martin. There are the islands of "Antigua, Montserrat, etc.", But they are all due east, I cannot go east. I am pissed off with myself. When the engine packed up, why didn't I accept the tow or just keep heading north, instead of wasting time trying to get back to Grenada. If I had, St Martin would not be so far away. Hindsight is a wonderful thing. Pity it does you no good. I console myself with the thought everything happens for a reason. If I had been in the Atlantic with this amount of water, it would be; "Game Over." With Mae becalmed, I had not bothered with my daily routine. Now she is moving again I will carry on with it.

Day 13

Once again, I must stop looking at the distance to St Martin counting down on the Gps; it depresses me. Mae has only travelled ninety miles. Water is a concern. Only having twenty-four gallons, I cut my intake to two litres a day. Another worry is I will lose the wind. As for food, I am eating canned soup and noodles which I cook together to save water and gas. The canned corn beef is inedible. It has turned into sludge due to the heat.

Eight pm I reinstate the routine hourly alarm call in case I sleep for too long. At nine I see strange lights in the water. Mae is cutting a swathe through them, leaving a dark patch with her wake. Once she has passed, the lights seemed to rush back in to fill it. Watching as they swirl about I think it might be phosphorescent plankton or small squid come up from the deep to feed. Turning to look forward I am startled to see a container ship heading straight for me, its navigation lights blazing. Mae's lights are on and I hope he can see them. So is the radio but I suspect the bloody thing isn't receiving or sending. I turn Mae to starboard. After ten minutes the aspect of his lights changes, he has done the same and we pass each other about a quarter of a mile apart.

His lights disappear over the horizon and I wonder if I should have sent up a flare. I reason that by the time he could have stopped or notified anyone of a boat in distress they would not find me. Anyway, I do not feel as if my life is in danger as long as I can make St Martin within the next three to four days

Day 14

Every new day brings my emotions full circle. Daylight changes my mood. Accepting my situation, I can get over any problems the day presents. But this will change when the sun sets.

The darkness will not only slide around me but also within me. I try to stop it but have realised it is futile. It is like the ebb and flow of the tide, inevitable, the way of the world I inhabit

Going through my morning routine is automatic and ends with me making tea. Then I settle down to wait for the batteries to charge, so I can turn on the Gps. Today, the heads-up shows me I have gained twenty miles on St Martin, speed is one knot.

Part of me wants to let out more sail and the wind to pick up. But not wanting the worry of the mast holding up under the increased pressure that this would create. I forget it and just leave Mae to plod on. At this speed I can put out a line and fish through the day. Catching two I grill them on the barbeque for dinner.

As I eat, the screen on the Gps beckons. I try not to look but I cannot help myself. The distance to St Martin ticks off oh so slowly, it's like watching paint dry. Over the last two days Mae's speed has not altered, and I have seen no other vessels since the container ship

Day 15

Woo Hoo! Mae made thirty miles during the night. The wind picked up and increased her speed. I am now only nineteen miles from St Martin and on a heading that will take me straight into Phillipsburg Bay. If the wind does not change, Mae holds her course, and with luck, I should anchor in the bay later today.

Everything I need is there. A marina, a boatyard, and marine engineering company. I do not want to count my chickens, but I feel optimistic. With Mae restocked with fresh food. Her water tank drained, cleaned and replenished. The chainplate welded and her mast reset. If they finish the repairs soon enough, I can still carry on across the Atlantic back to England. The wind drops off late morning. Mae is moving, just. So much for optimism. The wind stays light through the night and this morning. Mae's speed fluctuates between one and three knots. Progress is slow. Three pm. time for tea and a smoke. While I wait for the kettle to boil; I feel a change in the air. It becomes thick and oppressive; the temperature goes up, and the wind drops to a whisper. Going back to the cockpit I look around. Mae is wallowing, and in the distance, I see a murky haze surrounds us. It seems to reach up and merge with ominous dark clouds that sit above it. Except for the sound of the sea lapping against Mae's hull as she rocks from side to side. it is eerily quiet.

A storm is on its way, a big one! I am under no illusion that the bad weather I experienced before was nothing compared to what is coming. Mother Nature is giving me a grandstand seat and insisting I attend the performance. I am not sure how long before the show starts. So, I spend my time making sure everything that can, gets stowed away and secured. I especially take care of my water. I cannot afford to lose anymore. Also, I make a flask of coffee as I am sure I will not be able to leave the helm once the storm is upon me. By six pm I am only four miles out from Phillipsburg, willing Mae to get there before the storm hits

Visibility is bad. If it was clear, I might be able to see the island and some boat traffic, but I can make nothing out through the haze. Ahead the clouds have formed into thunder heads, blacker than the night. and the swell of the sea is steadily increasing

Bang on seven o'clock all the elements that have been cranking up come together and let loose with fury. It's my own piece of hell.

The wind switches from the east to northeast shrieking and howling like a banshee who wants to claim my soul. Mae's rigging sings out as if in agony. Walls of dark water rise and curl inwards threatening to engulf Mae. I fight to keep her heading, so she will take the waves on her port bow. *(I am terrified that if she gets hit amidships she will roll over.)* Mae is climbing the waves and sliding down the other side at frightening speeds Hitting the bottom of each deep trough with such force we both tremble, Mae with stress, me with fear.

Each time she remains upright ready for the next monster that will try to bury us. When we are in a trough, I can judge their height by comparing them to the mast. They are between fifteen to twenty feet plus and still rising. Deafening peals of thunder combined with the roaring sea blasts my ears, the drawn-out crackling of fork lightning speeds across the sky ripping through the clouds, adding to the cacophony. For a second there is a flash of light and I get a glimpse of the maelstrom that surrounds me. A creature from my worst nightmare sent to torment me.

Spray sweeps over Mae's deck, sea foam and sea-weed are getting caught on the rigging, then whipped away by the wind. It howls with rage because its trophy is not part of the boat. The sea is a seething angry mass that wants to drag Mae down to the depths and consume her. I focus on the bow. Looking up at the mast makes me feel sick. Not only because it is swinging from side to side, but I do not want to witness my half assed attempt to secure the forward chainplate give out and the mast collapsing.

My legs and shoulders are aching from the effort it takes to keep Mae from turning away and presenting her port side to the waves trying to kill us. I am barefoot, and my toes are sore and bleeding from trying to dig their way through the steel soleplate as I struggle to keep my balance. Counting the time between each wave, it is about nine seconds and I dread the prospect of one out of sync. If that happens it will break over us flooding the boat. As the hours wear on I become hysterical and make it personal. Laughing like a madman and screaming obscenities at the wind and sea.

"You fucked my boat, then becalmed me and now you want to drown me." Mae crashes down into a deep trough jarring my bones. The next wave towers above us. A monster, I shut my eyes. Mae climbs, and I yell. This is it! This is it! I am convinced she will roll, our demise imminent, she is heeling so far over. When she gets to the crest I open my eyes, she teeters fighting the forces trying to drag her back. Then skids down the other side hitting the bottom and sending a huge plume of water into the air.

Cackling like a lunatic I shout out. "You showed them Mae; You showed them. That the best you got you bastards." Each new wave I curse and fear. My rant carries on until I run out of breath and insults. I want to lie down, curl up and cry my eyes out, and I'm not ashamed to admit it. But I cannot give in to my fear. If Mae will not let the elements beat us, neither will I.

For ten hours we battle with the storm. It nearly breaks us but then I see light gather in the east. The sun rising, burning off the clouds, clearing them from the sky as it climbs. I stare at it muttering to myself and Mae. Greeting that beautiful yellow orb like an old friend and ally, willing it to hurry and kill my enemies.

Gradually the wind eases, and the waves lose their ferocious power.

Day 16

Seven thirty a m, the storm has passed, everything calms down, and it promises to be a fine day. From hanging on to Mae's wheel for twelve and a half hours my hands are stiff and cramped. I find it difficult to let go. Exhausted, but elated, we have survived. Still suffering from mental meltdown, I shout at the wind and sea, "You can't kill us you fuckers". As the adrenalin leaves my body, the madness fades, I look around the boat. Two battens are hanging out of the mainsail. It had been on the third reef. The headsail I had furled before the storm hit. So that was ok.

The cabin is a mess, things thrown about even though I had stowed them away. I have lost no water, which is a relief. The bilges have water in them but not enough to concern me. I will pump them out later in the day. More worrying is what will I find up at the bow after the punishment Mae has taken.

When I look at the ropes holding down the chainplate, some have worked loose. It's not a surprise that the split in the weld has travelled along a few centimetres more. But it held. How, I do not understand? After tightening the ropes, I go back to the cockpit. The mast will not collapse today at least.

I get out a hand mirror and look at myself. A madman stares back at me with sunken eyes sitting in black hollows. My hair has a yellow tinge about it, salt encrusted and frozen in place. Sticking out at all angles, and I have a scruffy dirty grey two weeks growth of beard. "Mae, we might not look pretty girl. But at least your afloat and I'm alive."

We had only been four miles from St Martin before the storm. Now visibility has improved I expect to see the island but there is nothing. The Gps is down. I am too tired to work out our position from the charts

While sitting in the cockpit I drift in and out of sleep. The last of my energy depleted. Like Mae I need to recharge my batteries.

It is now four in the afternoon and after dozing through the day I do not ache as much. But my mind is numb. The wind and sea have had the last laugh. Because of the storm and the change in wind direction it has driven us back and further west. Now sixteen and a half miles from St Martin and seven miles off my track into Phillipsburg.

The island of Anguilla may now be a better option. But, the Gps shows me it doesn't have the facilities needed to repair Mae. If so, I cannot see the point in going there?

I need to get back on course. Mae does not make it easy for me. After two unsuccessful attempts she throws a hissy fit and heels over so far that I am digging holes in the deck with my toes again. Once she is satisfied that I nearly shit myself, she comes upright and sails on. I do not understand what causes her to react like this. Whatever it is she does not like it and lets me know.

After what we had been through, I do not hold it against her, how can I? She kept me alive, I have an affinity with her. Since the storm I feel the bond between us is stronger. I know it's crazy; she is a structure of steel, an inanimate object. But I do not see her like that.

At around eight pm far off in the distance I can see the lights of a cruise ship. I guess it has come out of Phillipsburg and is heading for St Kitts or Nevis. As I watch it sliding across the horizon, I hear a low rumble of thunder. A shiver runs through me. There are more black thunderheads in the distance and the occasional flash of lightning erupts inside them. They are southeast of my position, heading south. So, I try not to worry. But after the last nights drama I can't help it. The cruise ship which is further east than me, has slowed to a crawl to give the storm time to move further south before continuing.

Even though I have left the radio on I believe it is faulty and useless. The last time I heard anything from it was when that ship was in trouble. That was two weeks ago, I nearly jump out of my skin when a voice speaks from it. Wondering if it is my imagination playing tricks on me, I wait. Then the voice speaks again.

My world has condensed down to just the wind and waves, Mae's creaking and groaning and flapping sails, and my voice as I chat/curse at Mae or the iPad. The sound of the voice throws me. I do not know how to react. It is not until another person replies do I believe it's genuine

Two Americans are speaking ship to ship on channel sixteen, the emergency frequency. Fascinated by this mundane conversation I go into the cabin to listen in. One of them wants to know when the cruise ship will start its firework display and if they are in a good position to see it. They waffle on for another couple of minutes till one of them says. "Hey, we better switch channels. Change to".

Losing interest, not wanting to hear any more about the poxy fireworks, I go back to the cockpit and look around. As I heard them so clearly, I think I might see them, or at the least their navigation lights. But all I can see are the lights of the cruise ship far off to the east.

It had not entered my head to speak to them, and my sub conscience asked me why? I answered like a petulant child. "Didn't want to." The truth is I didn't trust myself to say anything coherent.

Standing on the aft cabin roof and leaning against the mizzenmast, I stare at the cruise ship and wonder if I am too far away to see the firework display and do I care?

A bright light floods the cockpit. It's coming from behind me. Startled I grab the mast and swing around. Two spotlights blind me. When my vision clears, I can see a large yacht in full sail bearing down on me and closing fast. Thinking it is going to hit me I frantically wave my arms. The yacht hurtles past, ten metres off my port side. I cannot believe the figure at the helm did not see me as his lights swept over Mae. He didn't appear to even glance in my direction. Watching his stern light disappear I assume it's one of the guys I heard over the radio. Climbing back down into the cockpit I think. "The speed he is going, he must be in a hurry to see the fireworks?" Spending the rest of the night trying to manoeuvre Mae onto the track that will take me in to Phillipsburg is soul destroying. Once again, the wind is beating me. At one point during the night I am close enough to Anguilla to see lights and pick up a signal on my phone. Excited I send Mid a text. But it comes back, failed.

Day 17

I am not making any impression on St Martin or Anguilla. The weather is fine and the wind damn near non-existent, the wind meter reads force one. The waves a ripple. Great, if you have a cabin cruiser with bloody great engines. For a sailing yacht with no engine. "Fucking useless."

My resolve to get to St Martin under my own steam is evaporating. If I see another boat, I will ask for help.

The light is fading, and I have seen no boats or heard anyone on the radio. Towards St Martin, a glow is in the sky. It must come from there.

Day 18

It's three in the afternoon and I am at the end of my tether. Tired and frustrated, I can see the island but cannot get to it. The state Mae is in, she is just bobbing around going nowhere.

This galls me, we have come so far with no help and now under threat of falling at the last hurdle. After a lot of agonising, I send out a pan message which means, I am not in a life-threatening situation, but I need help.

I have no guarantee that the radio will send but I try anyway. Sending out the message every five minutes, on the third attempt I get a response. The signal is faint, I can just make out what the guy is saying. I Give my latitude and longitude and say I have no power. I also explain the radio seems to have an intermittent fault, he might lose the signal. He understands and says standby.

While waiting my emotions go into overdrive. Speaking to another person after so long puts me on a high, but this gets crushed by a sense of failure. By asking for help I feel I have let Mae, and Mid down.

It is taking so long for him to get back; I worry the radio will pack up before he does. After a ten-minute wait *(which seemed like hours)* the radio crackles into life. "The Dutch coastguard are on their way to you from St Martin. A cruise ship reported a yacht adrift and gave its position. It corresponded with the one you sent. That was a while ago, you have probably drifted further west." Time passes, there is no sign of the Dutch coastguard as the light fades. A concern is the coastguard will not tow me in and I will have to leave Mae to the mercy of the sea. I cannot afford to lose my boat. No matter what, I cannot abandon her.

The weather is changing. As the sun sets the wind freshens, and the sea becomes lumpy. The guy on the radio calls again. "Another boat is on its way to tow you." "Finding a boat and thinking it was you, they wasted time arguing with the skipper." "Convincing them he did not require aid has caused the delay." Sitting in the cockpit chain-smoking I stare into the darkness. Feeling numb, and indifferent to if I'm found tonight or tomorrow, such is my mood.

Nine pm, a flashing blue light appears in the distance, a searchlight is sweeping from left to right. Spotting Mae they head toward me. There are four guys in the rib. Coming alongside two of them jump onto Mae. This is no mean feat as the sea is choppy. Once aboard the rib backs off.

One speaks English. "If the towboat does not arrive soon, you have to leave the boat." "Sorry that is not an option, I will not abandon Mae." He tells me to get some things together anyway, just in case. I do as he wants, but I am not happy about it.

He asks where I have come from? After giving him a brief rundown of events. He says, "Why didn't you send up a flare or radio for help sooner?" "With the condition of your boat you were taking a risk." With a straight face I say, "Because it is Sunday, and didn't want to bother anybody." He repeats what I said to his colleague who looks at me in disbelief. "Your English?" I nod. They both laugh and shake their heads. "That explains it." "Do you have a long heavy rope to use as a bridle for towing?" "Yes." We set about rigging it up ready for when, if, the towboat arrives. With this done they go through the formalities of the paperwork for the boat and my passport. Satisfied everything is in order they wait until the towboat arrives.

They use this time trying to convince me to get on their rib and abandon Mae. I decline. It's ten o'clock when the boat gets here, and it takes half an hour to secure the tow rope. The skipper instructs me to keep the radio on and make sure Mae's rudder stays amidships.

While waiting for them the weather changes from force three. *(Gentle breeze)*, to force four. *(Moderate breeze)*. Wave height increases from one metre to two. This ride will not be smooth.

The two coastguards wish me luck, go back to the rib,

and leave. The towboat moves off, the slack on the rope tightens and snaps taut. Mae jerks then moves forward.

As speed increases Mae skews, first to port then to starboard. Like a fish caught on a line trying to throw the hook. Now and then crashing into a wave. The force sends a judder through the hull. Eventually she calms down and finds a rhythm, following the other boat like a dog on a leash.

After an hour standing at the helm I get tired. When I hear on the radio, ETA to St Martin is not for another hour and forty minutes, I think, "Mae's behaving, bollocks to this. I'm going below to make tea and roll a cigarette." Sitting next to the helm I fall asleep, only waking when I hear worried voices calling me. Still bleary eyed from sleep, the cigarette has burned down to my fingers, the mug of tea still in my hand, and not a drop spilt. I go out on deck and realise Mae is not bouncing around. She is being nudged towards a pontoon were two guys are waiting. They call out. "Throw us a line."

The skipper of the tow boat shouts down from his bridge. "Where were you? We thought you had fallen overboard?" When I say I had fallen asleep, he can't believe it. "Bloody hell, how? That was a rough ride." Thanking him and the crew for their help and taking a deep swallow I say. "How much do I owe?" *God knows how I will pay.* "Nothing, we are a charity, give money when you can".

He waves then turns the boat and moves off. The two men on the pontoon tie Mae's lines and come onboard. They are Customs officials. One says, "I can see you have had a rough time. We will not keep you long." "With your permission, my colleague will search the boat while I ask questions." I agree. Customs all over the world can do what they want when they want, they are just being polite.

Where have I sailed from? Where is my final destination? Is there drugs onboard? Am I carrying large amounts of currency? I answer all the questions and his partner finishes the search. He tells me to register in the morning. They wish me well and leave.

Too tired to take in my surroundings, as it is still early hours of the morning and dark, I go into the cabin fall on a bunk and sleep.

Chapter 4

St Martin

Day 19

I wake at eight a m, refreshed and eager to go ashore. I have a lot to do. My priority is to register with immigration. Then telephone Mid to let her know the situation and that I am ok.

The sights sound and smell of Phillipsburg hit me as I step out on deck. I take a three-sixty look around. Hills surround the bay. Here and there splashes of white, *(which I assume are villas)*, break up the colour of the land. Several yachts are at anchor in the bay. A distance behind me a huge cruise ship has just docked. Tourists eager to spend money hurry down the gangways. They look like ants scurrying along the boughs of a huge tree. Tied up on the other side of the pontoon next to Mae is a large catamaran. Twin hulls, white and pristine. Her stainless steel fittings gleaming. Smoked black glass curves around the centre cockpit. An easy two-million-dollar yacht.

I step off Mae and look at her from the jetty. In comparison Mae is a worse for wear shabby poor relation. But I wouldn't trade her for the plastic gin palace moored in the next berth.

A young guy walks along the pontoon toward me. Introducing himself as Simon the berthing master for the marina. "The coastguard left me a message they had placed your boat here last night." "The marina owner wants to know how long you intend to stay?" He is looking at Mae with a frown on his face. "It depends how long it takes to repair the boat." "Will that be a problem?" "I know I cannot remain moored here for nothing and will pay." When I mentioned paying the disapproving look evaporates. "I don't think so, what do you need?" "A mechanic to fix the engine and a welder to repair the bow chainplate." "Also, the water tank needs draining and cleaning." "It got contaminated." "What with?" "Diesel." "Once clean, I will need it refilled." He says, "There is an engineering company whose workshop is near the marina office." "They can do the work." He points to a standpipe along the jetty. "You can get water from there, when your tank is clean." "It has a meter attached, I will check with the office how much your mooring fees will be and add the water to your bill." "My office is that small building halfway along the jetty." "Come back there this afternoon and I will let you know how much your fees will be."

To save going to the main Customs and Immigration office which is where the cruise ships dock. He suggests waiting until ten thirty when the small sub office near the marina opens. It will save paying for a taxi. I thank him and set off for the marine workshop.

A tall Dutchman, "Eric" is the boss and owns the company. He will send a mechanic to Mae this morning to assess the engine. The welder cannot look at the damaged chain plate until tomorrow. "Come back at midday, by then I will have an idea as to the problem with the engine and an estimate of the cost."

I need to find somewhere with Wi-Fi. There are plenty of bars and cafes along the promenade. But being so early, few places are open. A shop called Candy Man, has an advert in the window. "Free use of Wi-Fi with every purchase." While at sea I suffered a craving for liquorice. His shop is full of it. Taking advantage of this I buy twenty sticks. The shop owner is a friendly and talkative American. He has lived here for several years and clues me up about the island. Giving me the code to access his Wi-Fi, he even boosts the signal for me.

Calling Mid on Skype I fill her in with what has happened. Since I left Grenada, she thought I was on my way back across the Atlantic. Relieved I am Ok, my safety is her main concern. She asks with the boat repaired, will it be too late to continue? The hurricane season starts on the first of June, which is today. She asks me to consider leaving Mae in St Martin and flying home. She also wants to know, "How much the repairs will cost?" "No idea, but I will let you know tomorrow, along with my decision." The battery on the tablet is low, so I ring off.

Thanking the store owner for the use of his internet connection, I ask him if he knows anywhere I can charge my laptop. The batteries on the boat can charge the tablet and phone, but not the laptop. I hope he might offer, but he doesn't. Instead, he advises me to go to a store on Front Street called "The Music Man." He is sure they will do it. There are two main streets in Phillipsburg. Front and Back. Front Street is full of designer shops and jewellers. Passengers from the cruise ships rarely stray from it, and the promenade. The lure of tax-free goods attracts them. Backstreet is more downmarket and mainly used by the locals.

The manager of "The Music Man" is helpful and warns me the electricity supply on the island is not up too much. The laptop will not achieve a full charge until the following day. I leave it with him and will pick it up tomorrow morning. A cafe called the Sinn Rose is near to the shop. There is still an hour until the customs and immigration office opens. So, I go there for breakfast and coffee.

When in a foreign country, I eat the local cuisine. But today I need a heart attack on a plate. It's on the menu. Eggs, bacon, sausage, and fried tomatoes. Two toast and a pot of tea. Chefs in Grenada do not know how to fry an egg. Everything else is ok but an egg, no. Thinking here would be the same, I was not expecting what was on the plate. The first proper fried egg since arriving in the Caribbean. With perfectly cooked crispy bacon, a fat pork sausage, and fresh fried tomatoes. After eighteen days of boat food, it is delicious. The best thing I had eaten while at sea, were the fish I caught while becalmed

After breakfast I still have time to kill before the Customs and immigration office opens. The video record of my journey so far is using a lot of memory on the tablet. Transferring them to an SD card will free it up. Inadvertently I delete a whole section. How? My emotional reaction is over the top. Distraught isn't the word for how I feel. They are the visual proof of my journey. Yes, I keep a log of events but it's not the same.

Pissed off with myself I go to the customs office to register. There are no problems, the process is over in minutes, unlike the crap I had to deal with in Grenada. Mae's previous owner had omitted to tell me he had not paid the cruising fees for a few years. After a lot of time and money, it got sorted, but it was an experience I would not like to repeat.

When I get back to Eric's workshop, my mood is still low over the deleted videos. Eric's news sends it spiralling down into my boots. "You need a new cylinder head. Yours has a split in it. There isn't any on the island, but I have located a reconditioned one." My mouth dry, I croak "How much?" He goes into mechanics speak. Still not giving me a figure. "Um, before I can tell you we have to solve another problem." "The owner of the marina wants your boat off his dock." "Why?" He shrugs. "I told him Hank my mechanic had to strip the old head down. So, your engine is in bits. We need the boat on the dock before we can proceed. But he didn't care. He has moved your boat and anchored it in the bay." I feel my brain melting and lose coherent thought, all I can say is. "Oh." Hank walks in and says. "Have you told him?" He looks at me. "Ah, I can see you have." Eric says. "I will talk to the owner tomorrow and try to sort this out. I do not understand why he has done this? Besides you, this affects me. If he turns boats away, it will affect my business." I fought down the anger building inside me and said. "I would understand if I was refusing to pay dock fees or if the boat was sinking. But I have not, and it isn't. What the fuck is the guys problem?"

How do I get out to Mae? Everything I own is on her and at the moment I do not have the cash to pay for a hotel. Besides, every penny I can get I need to pay for the repairs. Eric closes the workshop, and I join him and Hank for a beer. The bar is on the waterfront. On the beach in front of it fishermen are getting a small boat ready to go out. Hank suggests I ask them if they will give me a ride out to Mae. They are reluctant even when I offer to pay. The look of desperation on my face must have pricked the lead man's conscious. "We do not want your money. Be ready to leave in twenty minutes and you have your ride." I thank him and go back to the bar.

Eric had gone but Hank is still there. "Have you got a ride?" "Yes thanks, although they didn't seem keen at first." "What is it with people here?" Hank doesn't comment. Half an hour later I am standing on Mae's deck. Now what am I going to do? From being stranded ashore I am now stranded three hundred metres away from it. My mind can't cope with these latest problems thrown into my path. Laying on my bunk I fall asleep

Day 20

Last night, sleep had shut out the questions and frustrations of the day. Slamming a door on the screaming mob. Now I am awake they are hammering on it demanding entry. Refusing to think about any problems I ignore them, make tea, then sit on deck and enjoy the sunshine.

There is a medium size yacht anchored fifty metres away, a guy is loading gear into a small dinghy. Waving and calling out to get his attention. He acknowledges and signals he will come over when he finishes loading. As he starts the outboard, a woman comes out of the yacht's cabin and gets on the dinghy and they come across to Mae. The guy asks if I need help? Before I finish explaining my problem, he says. "We leave this time every morning. You can come ashore with us." "Are you ready to go now?"

Grabbing my rucksack and locking the cabin I get into the dinghy. As we make for shore, I introduce myself, and he tells me his name is Marco, and the woman is his girlfriend Sylvia. They are Italian, working in a store on the promenade selling Gelato. *(Italian Ice Cream)*. They have lived here on St Martin for the last eight months. Sylvia fly's back to Italy occasionally to see her parents. He asks how long I plan to stay? "As soon as someone repairs my boat I will leave." "Is there much work to do?" Sighing I say. "Enough." "Yes, we saw the damage." "Our boat had problems two years ago, while we were in Guadeloupe. It was a bad time." When we get to shore Sylvia asks if I would like coffee? How can I refuse? Marco insists I come back lunch time and try their Gelato. He and Sylvia make it fresh every day. Promising I will, I leave and go to the Sinn Rose café, and have a leisurely breakfast.

There is no point rushing to the workshop. Eric needs time to talk to the marina owner. Getting another coffee, I watch the world go by for an hour. When I get back to the workshop Eric says he will meet the marina owner this evening at six. He then tells me how much the reconditioned cylinder head will cost. Also, he gives me an estimate for other parts, labour, and welding. There is no way I can pay all of this myself. I must speak to Mid. Eric says until the guy who owns the marina allows Mae onto a mooring, he cannot proceed with the work. So, I have time to raise the money.

After collecting my laptop from the "Music Man" I go back to Marco and Sylvia and try their Gelato. It is tasty, and I ask for another bowl offering to pay, but they refuse to take my money. Marco tells me they will go back to their boat at five thirty. If I want a ride back to Mae I need to meet them then. Saying I will be back, I then go to the Candy mans to call Mid. "Have you decided what you will do?" "With the boat repaired will you still sail it back to the UK?" I tell her no. "If I can get the boat repaired I will take her back to Grenada." "Can you not leave it in St Martin?" "No, if a hurricane hits this place Mae would be at risk." "We could not afford to leave her on a secure mooring, and even if we could there is no guarantee, she would not get wrecked." She agrees with my plan but thinks it would be better if she pays for the repairs. As I will need the money I have to pay for the mooring and my flight from Grenada. We chat for a short while, then I go to a bar just to kill time.

At five thirty I meet Marco and Sylvia, they drop me off

and say they will pick me up early in the morning. Thanking them I then make tea and sit in Mae's cockpit wondering how Eric's meeting with the marina guy will go. It's a mystery why he removed Mae from the dock, but there is no point in worrying about it. What will be, will be?

Day 21

Not getting much sleep last night, this morning I feel irritable. I must get ready; Marco and Sylvia will be here soon. Ashore, we have coffee together, then I head to the workshop. Eric is there and has good news of sorts. He will move Mae tomorrow to a small wooden dock where the fishermen work on their boats. He will send Hank to pick up the reconditioned cylinder head later today, if I can pay for it now?

He lets me use the company phone to call Mid. I tell her what is happening and pass the phone to Eric. She pays for the cylinder head and money upfront towards the coming labour costs. He agreed she can pay using her bank card. This will save any delay to the work beginning as soon as Mae is on the dock. With the transaction over, he hands the phone back. Not wanting Eric adding the price of the call to the already mounting costs, I keep our conversation short, thank her and say I will try to call later this afternoon.

There is nothing more I can do, so I go to the Sinn Rose Cafe for breakfast. Conscious about money, limited in what I can spend and do. Boredom is my enemy here in St Martin. Not that there seems much on the Dutch side of the island. A lot of the bars close around five because most tourists are only here for the day and then go back to the cruise ships which then leave. The French half of the island may be different?

An older man and a young woman sit down at a table opposite. Calling over to a waitress, he has an unmistakable South London accent. "Here love, can we order breakfast?" We soon get talking and they invite me to sit with them.

His name is Ken, and the woman Sonja, his daughter.

They live in Brighton but spend the winter months in St Martin. Ken's wife was from St Kitts and Sonja's son lives there. They would like to stay and chat but must go to an appointment. They tell me the name of the bar they drink in and how to get to it. Arranging to meet tomorrow lunchtime for a beer. It doesn't matter where you are in the world, you always seem to meet someone from your home town. Leaving the café, I go to the "Candy Mans," call Mid then wander around the town killing time until I can get my ride back to Mae

Day 22

Marco and Sylvia arrive at the usual time and ask me to have coffee with them before they start work. Sylvia sits opposite me while Marco makes the coffee. She is a very attractive woman. Tall with a good figure, dark eyes, high cheekbones, and pouty lips. Her face framed by masses of curly black hair. As we drink, she studies me. "You are happier today?" "Work is to begin on your boat, is it not?" I laugh. "It shows does it?" "Yesterday you were, how do you say?" "Moody? Is that the right word?" Marco sits beside me and says. "It worried us." "We understand what it is like, stuck in a place you would rather not be." "It has happened to us a few times." "We are here and will help you if we can." Sylvia nods in agreement. "I have a few problems but now I think they are getting sorted." "Yesterday I let things get on top of me." "Today they will move my boat to a dock, so I will not need a ride back this evening."

Walking to the workshop, I think about how lucky I am their boat is close to mine, and to have met them. Without their help to get to shore every day, I would be in the shit. At the workshop Eric looks glum. "We can't move your boat until tomorrow." I know there is more to follow. "Also, the marina owner does not want you staying on the boat while it is on the dock. An insurance problem, but I think I can talk him around." "Some good news though, the cylinder head and the parts needed from the old one are here in the workshop." "Hank will prep it today and it will be ready to fit on Monday." Not trusting myself to comment I nod and say ok. Churning inside and feeling as if this guy who owns the marina is waging a vendetta against me. I walk away and go to the Candy Mans to call Mid.

She listens while I rant and rave about the marina owner. "Everything will work out, stop worrying." "The mechanics are working on the engine and if it comes to it, just stay in a bed-and-breakfast." "It will only be for a few days, I know money is tight, but it is what it is." Her matter-of-fact attitude and soothing voice help calm me down.

Ending the call, I walk back to see Marco to find out what time he will go back to his boat. "I am sorry our boss has invited Sylvia and I to his house." "We will stay the night." "We thought your boat was being put on the dock today, so you would not need a ride?"

I tell him it's not a problem, I will stay in a hotel or bed-and-breakfast. "You have a good time and I will see you tomorrow. I set off to look for somewhere to stay. It proves harder than I thought. A lot of the places I try are far too expensive or has no vacancies. On backstreet I find a place displaying a sign. "Room with an ensuite shower and toilet to let". A woman is sitting on the veranda and I ask her if it is still available. She says it is and leads me up a staircase to a balcony. There are four doors leading off it and she takes me to the last one.

When she opens the door, the Ritz it isn't, but will still cost one hundred dollars US for the night. I try haggling but she says she is only the housekeeper and must charge the price the owner sets. As I am paying with my bank card I cannot argue. Not being aware this morning about the problem with the boat I have nothing with me. After the woman gives me a key, I go to a store to buy jeans, T-shirt, and toiletries.

Back in the room I see someone has put fresh towels in the shower room and turned on the air-conditioning unit. It is making a strange clunking and grinding noise, struggling to cool the room. But it's better than nothing. Getting into the shower and feeling hot freshwater on my body is bliss. This is the first time I am free of salt since discovering diesel in Mae's water tank. Stepping out of the shower I catch sight of myself in the full-length mirror on the back of the door, and I'm horrified by what I see. Everyone has an image of self in their minds. What I see I do not recognise. A skinny old man staring back at me. His arms wasted flesh, and muscle. Loose skin hanging from his belly and stick thin legs. Instead of grey hair the sun has bleached it the colour of piss. It takes a while for it to register. It's me! When buying the clothes, I got the size I always do. Putting them on I look in the mirror again. They are hanging on me, but it's too late now they will have to do. I need to meet Ken and Sonja. The bar they use does not do food, so we drink beer and talk for a few hours. When they leave I am at a loss, what to do with myself? My room for the night has no tea or coffee making facilities and no TV, it is just for having a shower and sleeping in. It is too early to go back there and besides I am hungry.

I find a Chinese restaurant, with a bar. Four guys are sitting there talking to the barmaid, a young Chinese girl. I order a beer and ask for the menu, then sit at a table outside. While I'm sitting there one guy comes out for a cigarette and starts talking to me. He says I do not look like a tourist and asks me where I am from? When I tell him, I sailed here on my yacht and I am from England he invites me for a drink with him and his friends. His name is Franco and his friends are Julio, Mark, and Zaman. Franco is a pit boss in one of the casinos on the island. Julio sells coconuts. Mark deals in diamonds and Zaman lives by his wits. They ask me where my crew is? When I tell them, I am here on my own they want to know about my trip and where I plan to go next. I tell them that when I left Grenada, my intention had been to sail across the Atlantic but because of a few problems I finished up here. With my boat repaired, I will sail back to Grenada.

It impresses them, but they think I am mad. Even though

they live on an island, none of them has any desire to sail anywhere. If they did and could afford it, they will fly. To be in their company is great. I forget about my problems and enjoy myself. Franco tells me they come to this bar most nights of the week. They will be here Saturday and if I am free, I am welcome to join them. When I get back to the room, I am relaxed and slightly drunk. laying on the bed I crash out straightaway.

Day 23

The first job this morning is to see the housekeeper and book the room for another night. If Eric has no luck persuading the marina guy to let me stay on the boat, I will need somewhere to stay. Better to have it already in place. I find the woman sitting on the Veranda and ask. "Can I have the room for another night?" She says no. "You have no spare rooms at all?" She thinks, then says, "if you pay cash I can let you have a room for sixty dollars a night." "It is smaller than the other one and has no air conditioning but has a shower." "The owner has left and will not be back until the end of next week." "This will be a private deal between us, he does not have to know."

Before agreeing I ask to see it. The room is on the balcony at the top of the building. walking toward the door, I try to avoid treading on a few dead cockroaches. When she opens it, I'm sure I see a cockroach skitter across the floor and go under the bed. But the room is reasonably clean, and beggars can't be choosers. "As long as the cockroach doesn't bother me I won't bother it." If I can have it on the proviso of paying daily? I am not sure how long I will need it for, I will take it.

Deal done, I leave to get cash from the ATM and to see if they have moved Mae onto the dock. They have, and Hank is working on her engine. He has a message for me from Eric. "The owner has agreed you can stay on the boat." I know there's a but coming. "But, you can't until Monday." "You need access through our workshop, and when I finish

51

here, which will be soon, I have to lock it." "That's ok Hank, can I get things I need?" "Yeah sure, I'm sorry about this Roy."

On the way back, I go to the Candy Mans and call Mid. At the bed-and-breakfast I pay the woman for tonight then go to the Sin Rose for breakfast. The rest of the day I wander around the town stopping here and there for a coffee. Going back to the room at six pm for a shower and a change of clothes. I spend the evening in the Chinese restaurant. None of the lads from the night before are there. I have a meal and a couple of beers. Returning to the room at ten.

Day24

Waking at seven after a shower I go for a walk along the promenade while waiting for the Sinn Rose to open. I meet a middle-aged woman, who earns a living braiding tourists hair? As she is offering a service not begging for money, I let her put a small one in mine, while she works we talk. She is pleasant and well spoken, her name is Emily. She tells me about life here. When she finishes the braid, I ask how much I owe? She says two dollars; I give her five. The conversation about life in St Martin was worth that. After breakfast I wander around until seeing Ken and Sonia in a bar. I spend the afternoon with them. Then go back to the room for another shower before going for a meal.

Walking from the B&B towards the Chinese restaurant I bump into Emily. Recognising me she stops to talk. Her two daughters are with her. Emily says they are on their way to church. She asks, "would you like us to pray for you?" Thinking she means in church I say "Yes, thank you." I do not realise she meant right now. They gather around, and Emily asks God to bless my boat. "Please keep watch over this man and let no harm come to him as he travels across the sea." As she is praying her daughters are giving their assent as well. "We feel blessed you saw fit to bring this man into our lives and rejoice for this meeting." She then hugs me, and her daughters do the same. There is no guile, it did not seem false. A humbling moment. They wish me well and walk on. These people with nothing, well nothing much. Have everything that matters.

Living their lives and taking the knocks head-on. Nothing shakes their belief life is worth living. no matter what gets thrown at them. One of her daughters comes back. She lays her hands on my shoulders looks into my eyes and says. "Do not ask how, do not ask when, and do not ask why." "It just is, except it and you will be fine." Then runs off to catch up with her mother. Bewildered by this but feeling good, the next time I am whining about something I shall remember Emily and her wonderful family and shut up.

Getting to the restaurant I see Franco and the others are there. As soon as I walk in Alec (*The bar owner*) puts a beer in front of me. I say nothing about my encounter with the prayer group. It was a private moment. As the night goes on the beer flows. I ask Franco, "What do you do if a hurricane hits the island?" He says, "We live close together and help each other board up our houses. With our families gathered in one house, we sit it out. If we run out of water which is normally the case, we drink rum." "What about the children?" In a matter-of-fact voice he says, "you have to understand when we know a hurricane is coming we stock up with everything we can, but it is never enough." "A hurricane destroys all utilities." "The authorities always restore services to front Street and the businesses first." "People like us have to wait, sometimes weeks." The others nod in agreement. Julio says, "we have little industry or agriculture on the island." "The government earns its money from tourism." "Without the cruise ships stopping here our economy would suffer."

I say to them. "Here in St Martin and Granada I have noticed a wide gap between the indigenous and European populations of these two islands." "The wages paid to the local workers in the European and American businesses seem at subsistence level." "I had a chat with Richard a security guard in one of the designer shops near the Sinn

Rose Cafe, he has to work an eighty-hour week because of the low pay. Anything less he could not pay his bills and feed his family."

Franco who is a pit boss in one of the casinos and Mark the diamond dealer, get paid well, but amongst their friends they are the exception. This explains why they pay for the others round of drinks. As a coconut seller I can't imagine Julio earns much.

Zaman who is drunk gets off his barstool swaying and shouts. "Enough!" "You think this is bad, none of you know what hardship is." "My father is an Iranian, my mother a Jew." "Everyone in the world wants to kill me." Then laughing falls over. Franco picks him up and puts him back on the barstool. Julio says, "We need something to celebrate."

I tell them, the tenth of June is my partner Mids birthday. Franco gets Alex to line up shots along the bar. I show him how to take a photo with my tablet. We all stand up and raise our glasses and Alec takes the photo as we all wish Mid a happy birthday. Many more rounds of shots follow, and I get back to my room at two in the morning drunk but happy.

Day 25

Lifting my head from the pillow my forehead feels like an extended shelf. My tongue is stuck to the roof of my mouth. I need liquid. But there is no bottled water in the room, and I do not trust what comes out of the tap in the bathroom. A shower helps me feel better. Then I get dressed pay for the room for another night and go to the Sinn Rose.

After having a pint of iced water and juice, I order a light breakfast and coffee. Even though today is Sunday two cruise ships docked earlier this morning and disgorged their passengers. Crowds of people are swarming in and out of the designer stores opposite the cafe. Clutching tax-free goods. Hurrying to snap up what they perceive as must have bargains like sharks in a feeding frenzy. Before going back to the ships, which will leave in the evening and travel through the night to the next port of call.

Richard the security guard is on a break and joins me. Looking me over says. "Man, you look rough, out on the town last night?" He doesn't need an answer, the smell of rum oozes out of my pores along with the sweat. He gives me a cigarette and I buy him a coffee. "It's surprising how cheap booze is compared to food?" He grins and says. "Of Course, how else do you think they keep the natives from rebelling?" "Give them affordable booze, and they get so drunk they forget what to rebel about." "A cynical view, I know." "But I believe it to be true."

I buy more coffee and we smoke his cigarettes. He asks me, "when will you leave the island?" "Tuesday if they can finish the repairs by then." "Will you be glad to go?" "If I was a tourist with money to burn and time on my hands probably not. I would like to explore all the island. I have not been to the French side and would have liked to go there. But because of my circumstances I have to say yes." When Richard goes back to work, I take a walk along the promenade. Calling in on Marco and Sylvia for an hour and then go to the bar Ken and Sonja use. They are there and I while away the afternoon with them. Later going for a meal in the Chinese bar. Franco and his friends do not show. I presume they spend Sundays with their families. So, I go back to my room and have an early night

Day 26

After seeing Eric this morning I'm sitting outside the Sinn Rose having breakfast. The starter motor on the engine should be replaced. To do that he will have to order one and it will take a week to get here. I asked him if he could get a reconditioned one any quicker? The answer was no. Hank the mechanic said he can overhaul it but advises me to get a new one when I get to Grenada. Eric tells me the work on the engine and the welding will be finished midday Wednesday, two days' time. I bloody well hope so; I am haemorrhaging money I don't have on this island. One job I must forego is the emptying and cleaning of the water tank, I can't afford it. As I am returning to Grenada it should not be a big problem, hopefully? Passage time should be five to six days max. As a precaution I have brought another four, twenty litre Jerry jugs. With all them full there should be more than enough for the trip. Wednesday can't come quick enough. I've had enough of this island.

Chapter 5

Back to Grenada

Day 28

After two days of boredom and frustration. This morning, as I am leaving the room for good, I take my time having a shower. It will be awhile before I can have another. At nine o'clock I go to the Sinn Rose for breakfast. While I eat I can't help feeling like the condemned man having his last meal before execution. Now it is time to leave I'm not looking forward to it. I have had enough of this adventure for now. If I am honest with myself when I left Grenada, the boat wasn't ready, and neither was I. *(Not on my own anyway)*. When Mid suggested leaving Mae, I gave it serious thought. But again, my stubbornness took charge. We have invested a lot of time and money in Mae. We cannot afford to lose her. Besides getting on a plane and flying home would admit defeat and that would not sit well with me.

Buying supplies for Mae is easier said than done. There is a limited choice of fresh produce and the quality of what I find is questionable. Going to the dock I fill the jerry jugs with fresh water. They add the cost to my bill which after paying leaves my bank account empty. I still need to go to the main Customs and immigration office to pay for my clearance papers. Holding my breath when the official puts my debit card into her machine to take the money. If denied I will be well in the shit. To my relief it doesn't, and she gives me my documents. I have to leave St Martin by eight o'clock tonight. There is no point in hanging around and leaving in the dark, I will leave at five. This still gives me time for goodbyes.

Catching up with Ken and Sonja in their regular bar. We have a beer together and after exchanging email addresses and promising to keep in touch. I leave them and go to Marco and Sylvia's. I thank them for their help and support.

They tell me no problem. Marco insists I have a bowl of his Gelato before I leave. Once again, I exchange email addresses and promise to let them know I am ok when I reach Granada

Back on Mae I go through my checklist and make sure she is ready to go to sea. Marco and I had checked the weather forecast earlier. For the next two days the wind will be moderate *(eleven to sixteen knots)*, rising to a fresh breeze *(seventeen to twenty-one)*. I would prefer it to be light with a calm sea, as I plan to motor sail most of the way. With five hundred and sixty litres of diesel on board and the engine using about five litres an hour. I can afford to.

The engine coughs and splutters the first time I turn the key, second time it fires. When I increase the rpm black smoke belches from the exhaust. Checking the water pump is working, a dribble is being discharged but soon a healthy gush pours out. With the engine idling I get off Mae and release the mooring lines. Jumping back on I engage the drive and she moves away from the dock. By six p.m. St Martin is a couple of miles behind me. Hank advised me not to run the engine over two thousand rpm as it won't like it. He also said to give it a break, every three hours.

For now, it's at fifteen hundred rpm; I do not want to push it. Only the mainsail is up, more to help steady the boat. Mae's speed is only four knots, but I should still sight Nevis in the morning. The sea is running ragged, and the ride is uncomfortable. After nine days ashore, I am a little unsteady on my feet. But it won't take long to get my sea legs back. The Gps is showing I am three hundred and fifty-three nautical miles from the northern end of Grenada. At present speed I should make it in five to six days?

It's not long before the first problem. The autopilot has decided not to play ball and stop working. Lashing the wheel to keep Mae on course, I also set my alarm to go off every half an hour in case I drift off to sleep for too long

Day 29

Eleven a m: In the last eighteen hours Mae has put forty-eight miles between us and St Martin. Not a great performance but then it wasn't plain sailing. The sea was heavy. *(Still is)*. Our direction of travel, sea state and the wind are slowing progress. Following Hanks instruction to turn off the engine every three hours and leaving it off for two is not helping. It is running now but at one o'clock I must turn it off. Though I confess, I appreciate the relief from the noise. When it's time to restart it, a nagging fear plagues my mind. "will it?"

One p.m. Time to turn on the engine, it spins over but does not fire. I try to stay calm, but panic is hovering close by, and waiting to strike. With just the sails I can still head south but will drift west. Lifting the sole plate, I find the problem. The stop cable is not working. It's not returning the lever. When I push it back and try the engine again it starts. Before I can put the sole plate back in place Mae lurches. It falls onto my right foot and splits the nail on my big toe. The pain is excruciating but I soon get over it. The pain over the engine not starting would be worse.

Day 30

Seven a m: Another sixty-four miles is under our belt. Two hundred and eleven miles to go. I am not feeling well, the sea state hasn't changed, and the ride is grim. When the engine is off its worse. The wind dies down, but the sea stays the same. Lying on a beach in easy reach of a bar it would be a beautiful day. Out here on the boat is like sitting in a washing machine

Day 31

Nine a m. This morning has been a right laugh. I learn something every day about the joys of sailing solo. There are many types of waves that can hit your boat, a few are benign most are not. There are three that fuck my brain. I do not know the scientific names for them, but I have my own. The first one is the Sizzler, which runs along the side of the hull crackling and fizzing like electricity. The noise makes me freak out especially at night. They have always broke astern. Until early this morning. Two hit Mae in quick succession and broke against her sides. They were that high they poured over the dodger. They didn't damage the boat, only me psychologically. The next one is, "The Thumper." Not the small ones that tap on the hull. You get used to them. I mean the bastard's that wait until you're nodding off. They sound like the boat is being hit with a sledgehammer. The wave slams into the boat with such power. God knows how much force is behind them. But it's enough to make the boat rock.

The worse of the three is what I call, "The Dumper." He's a little gem.

Three hours ago, I made some liquid masquerading as tea. Coming from the galley I had to negotiate the steps up to the cockpit. As I did, I thought. "Bloody hell I haven't had a shit for three days?"

Mae on a starboard tack was healing over at a fierce angle, the sea running over the toe rail. In slow motion a wall of water rose above the aft cabin. Seeming to stop as if looking for a victim. The wave curled in on itself and roaring rushed towards me. Blowing out the canopy on its way into the cockpit and cabin. The drains in the cockpit could not cope so the bilge pump kicked in, I bailed like a madman.

That wave soaked everything in the cabin. My mobile phone on the table charging, I found floating on the floor, now useless. The Samsung tablet and Apple air in their waterproof bags had not suffered the fate of the mobile.

After the event I have no problem going to the toilet.

Someone once said. "The best pump on a boat is a frightened man with a bucket." How right that is.

I know sailing is a never-ending learning curve but doing it solo gets intense. I am constantly on edge, forever checking the boat. Twenty-four seven thinking, "what will go wrong next?" When something does, there is only me to deal with it. But there are moments that make it all worthwhile. I just wish they lasted longer.

As the day wears on the wind picks up and the once clear blue sky is filling with dirty grey clouds. I hope this isn't a sign of things to come. My biggest dread is the weather turning against me

Day 32

The sky cleared during the night. The threat of bad weather disappeared, and I had a panoramic view of the stars. Mae had eaten into the distance from Granada. We are now only one hundred and sixty-three miles away. Everything in my garden is looking rosy, until I go to start the engine. Nothing! Not a peep. I thought it might be the stop lever, but no. No matter what I do the engine will not start.

With just the sails I have the same problem as before. Mae will not sail close to the wind. Nowhere bloody near it. She will still head south but drift towards the west. The Gps is already showing we are fifteen miles off course, I can only hope when I am parallel with Granada I am not too far west?

Day 33

Disappointed with our performance last night. Mae only knocked twenty-five miles off the bill. There are large mats of Sargassum weed around. It's a nuisance if it wraps around the keel, rudder, or prop. So, I am avoiding them the best I can. There is a note on the map displayed on the Gps. "Do Not Enter This Area Under Any Circumstances!" We are skirting the outer edge of it. Maybe the warning is because of the weed. To the west I can see what looks like small islands of the stuff.

This afternoon the wind has slightly changed direction. I am moving away from the restricted area.

As the sun sets, I am clear of the weed, the wind drops

and is almost non-existent. As the sea calms down, speed is one knot. I am going nowhere fast.

Day 34

This morning everything has changed. The wind is back, and Mae is doing four knots. Today is one of those times this all seems worthwhile. The weather is fine, Mae is travelling in the right direction and I feel good. My one niggle is not having seen any other vessels since the engine packed up. I find it odd. Maybe I am further west than I think? Time will tell.

Day 35

Eight a m; Mae did well last night, we have seventy-four miles to go. Ahead an ominous line of black clouds is moving from the east toward the southwest. Tell-tale signs prompt me to reef in and secure the boat. Waves slap against the side of the hull. Now and then high enough to dump water into the cockpit. Nothing serious but enough for me to close off the canopy.

Sealed in the cockpit the air is stifling, but I will put up with it. The memory of the wave flooding the boat is still fresh in my mind. I don't want a repeat of that event.

One pm; the leading edge of the squall has reached me. It rains, as it hits the cockpit roof the noise is deafening. Wind whips around the boat and waves build in height and power. "Here I go again." Hanging onto the wheel while Mae bucks like a horse in a rodeo. This lasts for half an hour. The respite will be short, another one is racing towards me. The intensity of this second squall makes me think the first one was its baby brother. Before, Mae was taking the waves on the port bow. Riding up and along before sliding down the other side. Finishing in a good position to take on the next. Mae is now facing directly into the waves and crashing through them. A bone jarring experience. Twice Mae hits one that breaks over her. The roof of the wheelhouse is solid and with the canopy secure, little water enters the boat. This carries on for an hour then eases off

Checking our position, I find we are off course and further west. One hundred and four miles from Granada in the east. So much for my estimate of getting there in six days. Seventy-five miles to the west is the island of Margarita. I could get there easily. There is still enough water on board, but not for another eighteen-day stint. I will have to decide what to do? While I think about it, I will make tea and have something to eat. Five pm; I have made my decision. I will sail to Margarita. It used to be a popular destination, but from the things I have heard and read it would not be my first choice. This is out of necessity. There is no way I can face floating around going nowhere any longer. "Venezuela can't be that bad, can it?" "The whole population can't have gone rogue?" I will soon find out because that is where I'm going. I will take the reef out of the mainsail and fully pull out the headsail.

The mainsail won't move. Looking up the mast through my binoculars I can see one rope from the lazy Jacks has jammed itself into the groove the sliders travel along. If I want full sail, I will have to climb the mast and free it. Well what can I say? I am considering changing my name to Jonah.

After putting on a safety harness I go on deck but change my mind and return to the cockpit. Mae is still bouncing around too much. With no one on the helm to hold her steady I need to think about it.

I get Mae on a smoother course and lash the wheel. Turning on the camera of the Samsung tablet and setting it to record video. I place it against the windscreen of the wheelhouse focusing the view on the mast. My reason for this is if anything happens it will capture it on film. At least if anyone finds Mae there will be a record of what occurred. I'm not expecting anything to go wrong, it's only a precaution.

There are steps on the mast which makes things easier. Clipping the safety line to the highest step I can reach I start to climb. The only time I have been up a mast was on a small boat moored in a marina, not swaying around like this one. My progress is slow, this is more difficult than I expected. The higher I get the more the mast sways. Fuck it! The safety line stops me reaching the obstruction. Frustrated I hang on and look down. Even though I am nervous, the view of Mae and the sea below me is exhilarating. I wish I had a camera with me.

Returning to the deck a thought strikes me. "Connected to the safety line, if I lose my footing and fall?" "Yes, it would stop me hitting the deck or worse still falling overboard." But, "what if I hit my head and get knocked unconscious?" A vision appears in my mind of me hanging upside down swinging around with my head bashed in from hitting the mast. Unlikely but what if?

Even though I am tired and aching if I don't do it now I never will. Adrenaline pumping, I disconnect the line and climb back up the mast. Freeing the obstruction, I come back down fast. Then raise the sail to its full extent.

In the cockpit I let out the headsail halfway and adjust the main. The result is instant. Sailing downwind with more canvas out Mae's speed increases. Sitting down rolling a cigarette my hands shake. My subconscious reels off the *(things that could have happened to you, stupid bastard.)* Falling into the sea as Mae sails on, being the foremost.

Day 36

Checking the Gps, speed is six knots and we are heading straight for the north side of Margarita. When Mid cleared the cabin of the things, the previous owner had left onboard. She found an old pilotage book on Venezuela and had thrown it out saying. "Well, it's out of date and you're never going there anyway, so why keep it?" last night I found a book Mid missed, older than the other one. More of a cruisers guide than an aid for navigation. There are scraps of information about Venezuela and Margarita, but the book is twenty years out of date.

I pass Roca del Norte, *(North Rock)* and Los Frailes, *(an Archipelago of ten Rocky islands)*. Soon I will get my first view of Margarita.

A small open fishing boat powered by a big outboard with two guys in it cruise past. It is the first boat I've seen since the engine failed. I call out and wave. They glance in my direction but other than that ignore me

It would be nice to make Margarita before nightfall, but I know it will not happen. By eight pm the coast is three miles off Mae's port bow. Not wanting to get any closer in darkness I shorten sail and get enough course correction from Mae to run parallel with the island.

Ten pm; a boat with flashing blue lights travelling at speed comes towards me. A guy with a bull horn is standing on the bow. He shouts and from the way he is waving his arm I guess he wants me to get out of the way. I spin the helm to port and Mae turns but only slightly. But enough for whoever is steering the other boat to see it. He also turns and the guy on the helm stumbles but regains his balance. As the boat sweeps past Mae he shouts, and I guess curses me. The wash rocks Mae from side to side and now it's me stumbling about.

My radio is on and a stream of Spanish chatter is coming from it. Two different voices are speaking, and sound excited. There is more than one boat in the vicinity. Not understanding Spanish too well I do not know if any of this involves me. The speeding boat disappears into the darkness and the radio goes silent.

Half an hour later there is more excited voices and squawking coming from it. Amongst the Spanish I hear two English words. "Good job." At first, I think I miss heard, but then hear it again.

Another hour goes by, but the radio stays silent and I see no sign of the boat that passed me.

Whatever it was about, is over.

I wonder if they were on their way to rescue someone or were they chasing drug smugglers.

Hyped up I make coffee and smoke endless cigarettes

while keeping watch. But see nothing for the rest of the night.

Chapter 5

La Guardia

Day 37

Six a m; I am now past the headland and *(Punto Maria Libra)* and a huge bay called *(Ensenada La Guardia)* opens. To get to Pampatar. I must sail around the island. Not being able to face another night at sea I enter the bay and find a place to drop the anchor. a Venezuelan courtesy flag, and my yellow Q flag are flying from the halyard on my starboard spreader.

I had raised them the moment Mae passed into Venezuelan waters. The book warned of heavy fines if you did not observe this rule. Once anchored and flying the yellow flag, I am saying I'm not registered, if I meet anyone in authority at least they will see I am acknowledging their laws. Surely because of my circumstances. Engine failure, running low on fresh water, and exhausted, they will grant me some time. Is this a smart move or not?

The sea state is rough. It's not because of the weather; it is a nice calm day. I assume the configuration of the bay effects the water

I cannot make out many features along the coast, haze obscures it. On the starboard side there is a huge sandy beach with mangroves behind it. A stretch of water separates the beach from them.

As I creep further into the bay, the haze clears, and I can make out a white building in front of me. The nineteen eighty guide book I found on Mae says it is a Church. Through binoculars, it looks derelict. There are low-lying buildings off to the right. Behind and above the Church are small mountains. I am not sure what height defines a mountain, but they look too big to call hills.

I cannot see any sign of a town. As I get closer the wind changes direction. The mountains are affecting it and I have to change tack. Mae is now heading straight for the church.

67

A flash of light catches my attention. A minute later I see it again. Just to the right of the church partway up the mountain. Wondering if someone is signalling. I look through the binoculars again. Lack of sleep is taking its toll; impairing my judgement and rational thought. No one is signalling, the sun is reflecting on car windscreens. With the binoculars I can see a road running through a pass between the mountains.

Questions race around in my mind. The biggest and most worrying, "are the natives friendly?" Reason and logic says, they are. Venezuela is a civilized country. The narrative in the guide book describing La Guardia says. "A delightful church is close to the beach just outside a small friendly town. The town celebrate its patron saint each year on the twenty-fourth of May. Fishing is the main industry. Nice place to anchor and explore."

This conflicts with the odd scrap of news I had seen in papers and heard from people in Grenada and St Martin who had lived there for several years. They left because they did not feel safe anymore. "Venezuela is in meltdown," one couple had said. Soon I will find out. A small fishing boat with two young men in it pass by, heading towards the shore. Shouting and waving, I get their attention. The guy on the tiller turns the boat back towards Mae. His mate throttles back the engine, and they come alongside. Both study me with amusement. I must look like a wild man, with my shoulder length hair, and full beard, salt encrusted untidy and unkempt, shorts in tatters.

After a lot of hand waving and pidgin Spanish, I get them to understand I need help and a lift to shore. Agreeing to help they point to a spot closer inshore and tell me to move there and drop anchor. They wait while I change out of the rags I have on and lock the cabins. One of them helps me to board their boat. He can see the difficulty I am having.

Suffering from arthritis I should take medication every day but having run out of it days ago I am stiff and sore. Once seated, they gun the engine and head for the shore.

The two boys appear friendly, and we talk using sign

language and my limited Spanish. Their names are Juan and Henri. They are taking me to see someone they call, "El Torres". He will know what to do. Henri asks. Where have I sailed from?

I tell him "St Martin." "Your crew is on the boat?" "No, I am solo."

What would be the point of lying, I know I am taking a chance saying this, but I don't see I have a choice? Although impressed, he whistles through his teeth and shakes his head. As far as they are concerned this confirms I am loco. We go through an entrance to a small harbour. Large boulders laid out from the shore form a breakwater.

Small and medium sized fishing boats are moored up along with a large catamaran. A high wall runs along the shore side of the harbour with a large building behind it. Fishermen are working on their boats and nets. Some give me a cursory look but say nothing. I help moor the boat then follow the boys up the bank to a dusty causeway. Rickety huts line the other side. A tarmac road is at the end. We turn left onto it and walk about a hundred yards, stopping at a huge pair of steel gates. The boys argue about who will ring the bell. From what I can make of it neither one wants too. It is siesta and El Torres will be angry about being disturbed. Alex lost and rings the bell. Waiting for a few minutes he then rings it again. A gruff irritated voice comes from the intercom. Alex speaks so fast I can't follow what he says. Five minutes later, a man steps out and speaking perfect English with a Dutch accent introduces himself as Torres, and invites me in. Henri and Juan leave.

The gates hide a large compound with buildings either side and a wall at the end, which is the one I had seen from the harbour. In the middle is a swimming pool. We sit at a table in the shade and a woman brings two drinks. He listens while I explain my problem and is sympathetic and says he will help me. "It would not be wise to stay ashore at night." "You must go back to your boat." Someone will take you. "Have you lights on your vessel?" I nod and say yes. "Then I suggest leaving one on during the night to discourage visitors." "Your boat is on a path that some unpleasant people use at night." He laid a finger alongside his nose. "You understand?" "Venezuela is a dangerous place, you must leave as soon as you can." Leaving me sitting at the table he walks over to the wall overlooking the harbour and calls out to someone. After a short conversation with whoever is on the other side, he waves me over.

There is a man standing below the wall. "This is Silvio". "Go with him he will see you get back to your boat." "Leave now you need to be on it before nightfall." "Silvio will arrange with you to return here tomorrow." He is friendly and speaks English, he gets me a ride on a fishing boat going out to sea for the night. he will come out to Mae early in the morning and get me back to Torres. "Do not worry, you will be ok."

Ten minutes later I'm back on Mae alone with my thoughts. Sitting here in the cockpit with both cabins locked. writing in my journal. Listening for the sound of approaching boats. If I hear any, I hope it is returning fisherman. Not the bandido's, drug runners Torres warned me about. He had said word travels fast here. A foreign yacht with just one gringo on board might be too much of a temptation to resist. Easy to rob and strip the boat of equipment.

They will get a surprise if they come, I have the flare gun loaded and a cutlass close to hand. "Cut through bone no problem," Glenroy the diver in Grenada had told me. The best weapon I have is the will to use them, I have not been through all this grief to give everything away. After the last thirty-six days I thought things could not get any worse. How wrong can you be? I have a restless night, spending most of it in a half sleep state. My nerves wound up tight like piano wire. The slightest noise, which there are many, bring me wide awake.

Day 38

At two a m; I hear the unmistakable sound of an outboard engine approaching Mae's starboard side. I grab the flare gun pulling back the side of the canopy and keep low, so my silhouette will not show up against the red nightlight. Looking out a small fishing boat is coming towards Mae with three men onboard. Standing up and raising my arms as if stretching I spot the boat. then sit back down. They must have seen me because they veer off and head for the shore. Having no idea if they intended to board Mae or were curious. Either way, I am glad I did not find out.

Too much adrenaline is running around my body to sleep. My heart is pumping, I need a cup of tea and a smoke to help calm down. If I ever run out of either *(especially the cigarettes)* any pirates will have a crazed nicotine junkie suffering from withdrawal to deal with.

Six thirty a m; Silvio paddles towards Mae in a green kayak. He pulls alongside; he has brought slices of melon, bread, tuna, and peppers. Also, a large thermos of coffee. Inviting him aboard we sit and talk while I eat. Silvio's English is as good as Torres's, so there is no problem communicating. Silvio has a laid-back attitude and wants to help.

He had seen Mae yesterday tacking back and forth in the bay and guessed I was in trouble, why else would I be here? He travels around the Caribbean on a long-range fishing boat and occasionally has problems. Help always comes from other sailors, he knew how I was feeling. So, thought it right he should try to help me.

He said. "Venezuela and her islands have long since stopped being friendly to tourists, especially Americans and most Europeans." "Torres told me yesterday, Venezuela is a dangerous place and I need to leave as soon as the boat is able." "He is right," said Silvio. This country's being run by a corrupt government who are turning everything to shit.". "Incidents of robbery, kidnapping and murder increase daily. There is little food in the stores. Unless you are in the military or a government official, you have a hard time trying to buy any. The officials and army come first. Ordinary people with families get left with the dregs. We must turn to the black market for the most basic things. I am sorry you see my country now. Torres and I will do what we can to get you away from here. For sure someone will report your yacht is in the bay to the authorities. If they find you here?" He looked away and shook his head. "What?" "Believe me you will not have a good time." I didn't know what to say. Silvio had said it so matter of fact and with such feeling.

I told him that during the night, a fishing boat with three men in it had approached Mae but had veered off when they realised I had spotted them. Silvio said. "Not everyone here is bad. Having lived here all my life I know most of the fishermen. The majority, decent human beings who would not harm you, but there is a few I do not trust. My advice is always being aware of who and what is around you."

He asked if I had any weapons on the boat I could use for defence. I produce two machetes, a diver's knife, and a flare pistol. Frowning he says. "This is all you have?" I could tell it did not impress him.

We talk for another hour sharing information about ourselves and how different our worlds are. He calls out to a passing fishing boat heading back to the little harbour. The old guy in the boat knows him, he waves and steers the boat towards us. Silvio asks if he will ferry me to the harbour. He agrees and brings his boat alongside Mae. The new cranking battery purchased from Eric in Phillipsburg needs checking. One of the Torres's men will do it. As I pass it over to the old guy, he says something to Silvio. I catch the words, "Gringo and En Venda." Silvio replies and they both laugh.

He sees me looking at them with a frown. Silvio says, "Rico asked if you want to sell it?" "He needs one, batteries are scarce. I told him he could not afford it."

Holding up a hand showing to the old fella I want him to wait, I get back on Mae. There is the old battery on board. It works fine but is two years old, I was keeping it as a spare.

Passing it over to Rico I ask Silvio to tell him he can have it. Once he understands he tries to refuse but I insist. Through Silvio he tells me thank you and appreciates my generous gift.

Silvio tells me "When you get ashore go straight to Torres, do not hang about. People know I am helping you and you are under the Torres's care, so you should be all right. But please go straight there! I will see you sometime tomorrow amigo." He thanked the fisherman then paddled off heading towards a distant beach.

I try to talk with the old fisherman but with my limited Spanish and his non-existent English it is hard work. we both use a lot of sign language. By the time it takes to get to the harbour, he gets it that my yacht needs repairing, and I understand after fishing all night there is not much to show for he's efforts.

I help him moor his boat and offer him two dollars for the ride. He looks at me offended and protests that payment is not necessary and refuses to take it, but I get him to accept.

As I walk along the causeway that runs along the top of the harbour, there are a few people around. A couple give a friendly smile and say. "Hola, Buenos dias." Others look at me with suspicion turn away and stay silent.

After five minutes of walking I am outside Torres's big steel electric gates, sweat dripping off me and my arms ache from carrying the battery. The temperature is already thirty-four degrees, and it is only nine a m;

Pressing the entry button, I look up at the cctv camera covering the entrance. After waiting a few minutes, I hear bolts being drawn on the door set into the gates. A man beckons me inside, shuts the door and locks it, then takes me to Torres's private quarters. He is sitting at a table drinking coffee. "Good morning Roy. You are well?" "Good morning Torres. Yes, thank you."

He tells me to give the battery to the man who brought me in, then waves him away. "Pedro will see to your battery, although today being Saturday, you will not get it back till

Monday or Tuesday. I have arranged for someone to come out to your boat later." "He works for me from time to time." "He will try to fix your engine." "He is only young but knows a little about them." "If he cannot do it, you will wait until Monday, when the mechanic from town comes back from the mainland."

I thank him and ask, if I can buy cigarettes from a shack I saw down the road, it appeared to be a shop. He looks at me as if I am a wayward child and sighs. "No, you are a foreigner carrying a rucksack, they will kill you for it, even if there is nothing in it." From his attitude I can see there is no point in arguing. The conversations with this man are always short and to the point, He is used to giving orders and being obeyed. Not that I feel he is giving me an order. He knows the way it is here, I don't. So, I say to him. "I will take your advice." "Good, are you hungry?" "Eh, yes." "Get yourself some water from the outside kitchen in the compound, it's on the left passed the pool. I will prepare breakfast."

Then he left the room. I go out to the kitchen. There is a couple sitting at a table, unlike the fishermen and the few women I had seen on my walk to the Torres's, the man's clothes look expensive. He's partner is slim and attractive wearing quality clothes.

They are chatting, drinking coffee, and smoking. He looks up at me and says. "Hola Anglaise, El Torres told us about you, I am Carlos, and this is my wife Mita. I introduce myself and we shake hands. I ask him. "Having a good time? "Yes, Venezuela is a beautiful country, Margarita is a wonderful place, even with the troubles. "We are from Madrid, I am here on business and staying with my old friend Torres? Please excuse Mita she doesn't understand English."

What do you think of my friend El Torres?" Before I can answer he says. "He is, how you say in English, abrupt, yes?" "But do not worry." "If he says he will help he will." "Do you need anything?" "Yes, cigarettes, but Torres told me I cannot go to get them myself." "No, he is correct, it

74

would be most unwise, at the least they wouldn't sell you any." "We are going out, I will get them for you if you wish?" "Thanks, I would appreciate it." "Do you have any money?" only American dollars. "Ah, that is a problem, you need bolivar's, you cannot spend American dollars." "Give me a couple and I will buy cigarettes for you out of the bolivar's I have."

He hands me a half-empty pack. "Take these, they will last you until we return." I thank Carlos again and walked back to Torres's quarters, he is sitting at the table, on it a small plate of fish and a baguette. "Eat and then you can use my Wi-Fi to talk to your wife." He slides a piece of paper towards me. "This is the pass code." "I change it every day, for security reasons."

When I get through to Mid, Torres leaves saying, "I have things to do." Aware of the time I keep the call short. I don't want him to think I am taking liberties. I tell Mid I am in Venezuela, and not to worry. She is now very concerned for my welfare. Reassuring her I am ok, and that people are helping me fix the boat, we will speak soon.

When I finish I go back outside, Carlos and Mita are back. "Sit down, Mita will make coffee." He gives me four packs of cigarettes. We sit and talk. Mita brings the coffee then sits with us, a bored look on her face.

Torres comes and tells me join him at the wall that looks down on the harbour. He calls out to a guy working on a boat and tells him to take me back to Mae. The man says something in rapid Spanish and holds up his hands, his fingers spread out. "He will take you in ten minutes." I ask if I can come back tomorrow to call Mid. "Yes of course".

He repeats his speech he gave me yesterday about leaving Mae unattended then looks me straight in the eye." It is very fortunate they brought you to me first." I assure him he has made me aware of my situation. Adding, "I always seem to meet the right people at the right time. I think someone is sitting on my shoulder looking out for me and I don't know why?" "Maybe it's because you earned it Roy?" I must have a puzzled look, he pats me on the shoulder. "Do not worry about it now gringo, you must go, I will see you tomorrow.

Back on Mae, I wonder why he had said that. Late afternoon a small boat pulls alongside with two young boys in it. The older one points to himself and says "Hector", then he points to the harbour. "El Torres." He is the kid sent to look at the engine. Throwing me a rope, I catch it and pull the boat close enough for them to board Mae.

He takes the rope off me and lets out enough so their boat drifts away from Mae, then ties it off. Once I have lifted the sole plate above the engine and given him my tool kit, he doesn't hang about. He gets down into the compartment and goes to work on the starter motor, the younger boy passes Hector any tools he asks for.

It is hot in the cockpit and the humidity is high. The engine compartment is like an oven. Sweat rolls off Hector and he has to stop every ten minutes and come up for air and a drink. I offer them orange squash, but they prefer plain water.

Hector has a smattering of English and during the short breaks, I learn they are brothers, the younger boys name is Miquel, he's thirteen and Hector is seventeen. Hector works with his father and uncle on a fishing boat, but he really wants to be a mechanic. Knowing enough to do simple jobs by watching and working with the mechanic in the town. But working long hours with his father leaves him with little spare time.

He does not want to stay in La Guardia doing the odd repair on the fishermen's outboards. To be a proper mechanic, he must go to college. His father cannot afford to pay. "I am saving from the little I earn." "I will probably be an old man by the time I have enough."

He is having trouble getting at the last bolt holding the starter motor and the sea is getting choppy. With Mae rolling I can see Hector is feeling sick from the motion combined with the smell of diesel and oil. I know how he feels, being in that confined space at sea is no fun. I tell him to stop and leave it for tomorrow.

Day 39

Every small sound last night made me nervous and had me reaching for the flare gun and machete. It is eight a m. Silvio still isn't here. There is no activity onshore that I can see. Sitting here with no way of contacting anyone I feel marooned, isolated from the world. At sea, I do not feel this so much; I am occupied, still heading towards home. Even if the distance I travel is minimal.

It's hard to describe. Being stationary and just waiting with the means to contact Mid less than a mile away, is awful. There is nothing she can do. but talking to her even for a few minutes helps.

The men helping me are under no obligation to do so. They have their own family's and lives. "You expect too much," I tell myself I am way down their list of priorities.

At ten a m; there is still no Silvio, maybe because it's Sunday he cannot come. By one pm Silvio still has not shown up, I can't help wondering why? Two boats leave the harbour and head towards me. One has an old guy at the helm, two young kids are sitting on the back. They can't be over five or six years old. None of them wearing life jackets or harnesses. He circles around Mae pointing at her.

The children look excited at seeing the yacht close-up and wave. I wave back, and they shout out "Hola Anglaise." The boat speeds up, so the bow is high and the stern low in the water. The kids scream and laugh enjoying the thrill as they head back towards the shore. Health and safety in England would have a fit.

Hector, his brother and two other fishermen are on the other boat *(one of them is Hectors father)* They pull alongside Mae. After exchanging greetings Hector wants to know if he can borrow a pair of snorkelling glasses. He had seen I have two pairs, when on the boat yesterday. They are going spear fishing.

I hand him the glasses, he takes off the snorkel and gives it back saying he doesn't need it, promising to bring them back later in the day. I tell him no problem, mentioning that I have not seen Silvio. "Silvio is with his family but will be out to see you tomorrow."

So, I am stuck on Mae and cannot contact Mid until Monday. For the next two hours I keep myself busy cleaning and tidying the main cabin. The sound of an outboard engine close by sends me scurrying back up to the cockpit. Hector is back, and I can see that his father is tipsy.

With a beaming smile Hector's dad passes me a bag filled with fresh mussels, lying on top are two fat fish. Hector tells me to fill a bucket with seawater and put the muscles and fish into it. They will keep for a while. When he hands back the dive glasses, I refuse them. "Keep them they are yours." His father asked him a question, I do not understand what he says. Then Hector asks me. "Would you like to go to El Torres for a while and we will return you to Mae later?"

With assurances they will meet me in the harbour at seven pm and not let me down. I lock the cabins and climb aboard their boat. When I arrive at Torres's there is a party going on.

Apologising for gate-crashing, Torres says, "You are not, your welcome here anytime". He knows I want to call Mid. So, he gives me the Wi-Fi code of the day and lets me use his dining room, so I have privacy. When I finish talking to Mid, I go out into the compound. There is a barbecue set up, and I am encouraged to help myself to the food. People are friendly, but conversation is difficult. None of his guests speak English. But I enjoy the company and time passes quickly.

Thanking Torres for his hospitality I go down to the harbour to meet Hector. It's now dark and seven o'clock passes by. With no sign of Hector or his father I begin to worry. Because I am only the other side of the wall that separates Torres's property from the harbour. The Wi-Fi signal is still available, so I call Mid again. She is concerned. "What will you do if you can't get back to Mae tonight?" "Go back to Torres's?" Before I can answer

Hector and his father's friend shows up. It relieves Mid when I tell her they are here and will call her tomorrow.

While I wait as they get the boat ready a drunk guy walks up to Hector and shouts at him. He is looking me up and down with distaste. He makes it obvious he does not like me and spoiling for a fight. Hector ignores him and when I ask if there's a problem? He says. "No, get into the boat." Some huge fellow looms up behind the guy making a fuss and brings one of his fists down on the top of his head. The drunk crumples like a sack of shit. When we are all onboard, he pushes the boat away from the dock. As we leave Hector calls out to him. "Gracias, Jorge," he waves then walks away, leaving the drunk laying on the ground. Instead of heading out of the entrance towards Mae we cruise further along the harbour and come to a stop opposite a shack. It's the fisherman's bar, lively Latino music blasting out from the open door and windows.

People are outside. Some sitting, some dancing, or just standing about. Most are drunk. For one uncomfortable moment I think Hector wants me to go for a drink. But I needn't have worried. We were there to pick up his dad.

Hector's father comes stumbling towards the boat with a bag in one hand and a bottle in the other. He clambers into the boat and we take off for the entrance. With the boat bouncing around Hector's father stands up on the centre seat and swigs from the bottle and starts singing. No one seems worried that he could fall out of the boat.

When we reach Mae, from the bag Hector's father presents me with a bowl of hot steaming mussels. Insisting I try them and have a drink with him before I get off their boat. They are delicious, but I tell him I don't think the slug of fire water enhances their flavour. Hector translates for me. The man roars with laughter and tells me "Your, Ok." Come fishing with us Wednesday. Promising I would, I climb aboard Mae. Hector passes me up the bowl of mussels and they wish me good night. As I sit in Mae's cockpit eating the mussels, I watch them as they head back to the harbour. Hectors dad still balancing on the seat singing his head off.

Chapter 6

"You Speaka de Spaaneesh"

Day 40

Mae's slow rise and fall, waves lapping against the hull, and chirping of the cicadas on shore was my lullaby. My mind, racing with cluttered thoughts; eased of the gas and I drifted into a fitful sleep. I am not sure if the dawn breaking or a premonition woke me.

It did not take a lot of working out that something was wrong. Mae is wallowing and jerking on the anchor chain, like a boat tied up on a mooring in a high wind, snatching at her lines. Going up on deck and I see Mae has moved further down the beach from the little fisherman's harbour. The anchor dragged during the night.

The mechanic Torres has arranged will arrive today, he will not appreciate having to come out this far to the boat. I make breakfast. Tea with a small bowl of melon. Food is at a premium and amazingly there is still gas in the cylinder. It's only five o'clock in the morning which helps. I have to haul up one-hundred feet of chain plus the anchor by hand. The temperature is already twenty-seven degrees, in a few more hours it will be up to thirty-one.

I suppose because of the time spent in the Caribbean I have become accustomed to the heat. If the temperature drops below twenty-six overnight, I feel cold.

I pull up the anchor then unfurl the mainsail and head out into the bay. There's not much wind, it takes time to get the boat moving. Once far enough out I turn back towards the shore and try to line up the bow on the small harbour.

As usual Mae is being a bitch, and I must tack backwards and forwards a few times to get her even close to where I want to be. Concentrating hard, I do not hear the boat that has come up on the stern until it is within a couple of feet.

Turning around expecting to see a fishing boat on its way back to the harbour after a night fishing, my heart skips a beat. It's a military patrol boat, bigger than a fishing boat and painted in a dark green and grey camouflage livery. Aboard are three soldiers in dark green uniforms. Two are holding automatic weapons. The third is steering the boat, he has a pistol on his hip.

The soldier standing at the bow shouts in Spanish and using hand signals shows I should stop what I am doing and drop anchor. Realising I do not understand, he raises his rifle and reverts to pidgin English. "Venezuelan police, Marine commando, stop now! I come aboard".

Trying to make him understand I need to drop the sails is not getting through to him. He keeps shouting and making threatening gestures with his rifle until I go on deck to the anchor.

Everything Silvio and Torres told me about the Venezuelan military churns in my mind, my stomach joins in. I want to be sick. Dropping anchor, then going to the mast I lower the mainsail and hurry back to the cockpit and deal with the genoa. The one who shouted at me climbs aboard. The rifle slung across his shoulders and wearing a cheesy grin. Somehow, I smile, there is no way I will let this guy know I'm shitting a brick. He is anxious but with an automatic rifle and backup not as anxious as me.

There's a Chinese look about him *(I learn his nickname is Soto)*. "You speaka de Spaaneesh." He said. "No, I speak English". Cocking his head to one side he repeats the question. I give him the same answer. He looks across to his mates and in Spanish rattled off a question to them. One of them gets out a mobile phone the other trained his rifle on me.

The guy onboard says. "Papers, passport." Miming that I must go below into the cabin to get them. He pops the magazine from the rifle inspects it and smacks it back down to make sure it was in. He sees me looking. "You, Speaka de Spaaneesh." I try a different tack and point to myself. "No, no Spanish. Anglaise, London."

He motions with the rifle towards the open companionway. "You go, you go, get papers". Following me down into the cabin he watches while I open the drawer which has my documents. The barrel of the rifle is pointing at me always. He looks around the cabin with interest. I point to the kettle standing on the gas hob and ask if I can make tea, and would he and his friends like some? He still has a smarmy grin on his face, wrinkles his nose and shrugs. I pantomimed making tea. Then he understands. Wanting none for himself or his friends but agrees I could make one for myself.

Once I make the tea, we go back up to the cockpit where he takes my papers off me. Pointing with the rifle for me to sit down by the helm, he sits at the back of the cockpit and rests his rifle across his knees and flicks through my paperwork. As I sip my tea. I can tell he did not understand what he is looking at regarding the ship's papers. Putting my tea down I pantomime having a cigarette. He nods, and I take it as a yes. When I pick up my tobacco, before I can roll one he snatches it away from me. Jumping up, he gets excited and shouts. "Marijuana, marijuana." "No!" "tobacco." He holds it to his nose. "Not marijuana?" "No, tobacco" "It marijuana." "You Americano Gangster, Chicago." He waves the pouch at his friends. "Marijuana, marijuana," he shouts, and they all laugh.

They have an excited conversation. Grinning like a chimp, he hands me back the tobacco pouch and says "make," miming rolling a cigarette. I do as he asks and hand him the first one. He lights it, takes a deep drag then pulls a face. "Not marijuana." I shrug and say, "No, tobacco." Disappointed he throws it over the side. The other two lose interest and do not want one after that. So, I roll one for myself sit back drink my tea and smoke it. Thinking. "Oh great!" Boarded by armed morons. The tune of duelling banjos from the film deliverance plays in my head

Going through my paperwork again, he opens my passport. flicking through it. I'm getting agitated at the way he is handling it and worried that he would rip the pages. I ask him if he could be more careful with it. He grunts and then gets excited when he finds the pages with the entry stamps. He stands up and shakes the passport at me.

"Venezuelan no stamp! Venezuelan no stamp!"

Standing up trying to explain to him why, the pretence of friendliness evaporates. He brings up his rifle, the muzzle pointing at my chest. I hear the snick of the safety as he thumbs it off. "You sit down, your American gangster, Chicago".

He calls to his mates. The one with the phone talks into it and the atmosphere changes. As I sit back down, he goes to the back of the cockpit and sits on the aft cabin roof. His smile is back on his face, but I can tell by his eyes if I move, he will not hesitate to shoot me.

His voice takes on a threatening tone. "You speaka de Spaneesh gringo. Your American gangster, Miami. Again, I try to explain I am English and why I am in the bay. He is not listening. My mouth dries out through fear, but I do not show it. Its bravado, to show fear will probably make matters worse for me. How much worse I do not want to think about. I ask him if I can get a drink of water? No, is the curt reply. After a few minutes the guy using the phone calls over, he has instructions from whoever he spoke to. They put out fenders and bring their boat next to Mae and tie off on one of her cleats.

They have a long conversation amongst themselves which seems more like an argument what with all the arm waving and shouting. I do not understand what they are saying, but I get that the guy with the phone is in charge or supposed to be? He is having difficulty in getting Soto to accept his orders. He acts like a petulant child, but eventually sits down and sulks. The other two make themselves comfortable in their boat but keep their rifles pointing in my direction. Sitting in the heat sweat is running down my chest and back. It doesn't seem to bother them, even though they are in full uniform. There is no conversation. I assume we are waiting for someone or something to arrive. After what seems like an age another message comes through on the phone. They tell Soto to search the boat. We go down into the cabin. While he searches the drawers and cupboards he orders me to stand back. He is not being thorough, just a cursory look and see. Finding my Samsung tablet, he does not understand what it is. I try to explain and show him how it works. He likes the fishing rods hanging up inside the cabin. Taking one down he brings it back up to the cockpit telling me to follow him and bring the tablet. I show him how to take a photograph with the tablet. He gets excited and against my better judgement, worrying he will drop it into the water. I give him the tablet.

He takes a photograph of me then hands the tablet over to his companions, so they can see his handiwork. They are none too careful with the way they handle it. They had no desire to have photos taken of themselves. It surprises me they hand it back I thought they would keep it.

I then show him how to use the fishing rod. Casting a line out he laughs but soon gets bored and hands the rod back. This activity eases the tension and now the atmosphere is slightly more relaxed.

Soto says they are my brothers, amigos, and I am not to worry. Then looks over to his mates in the other boat and laughs. I know he is taking the piss and getting on my nerves. I want to say to him. "You are not my friend and you are not my brother, you're a fucking arsehole. Instead, I ask if I can get a drink of water, he waves his hand and says, "Si`". I do not bother to ask if they want any. I sit down roll a cigarette and light it up. He glances across at me but makes no comment, he is copying down details from my documents onto a piece of paper. Now and again he beckons me over and points to a word in the documents and asks me to explain what it means. This is all very difficult due to the language barrier. When he finishes, he lies back and watches me.

The whole situation is surreal, my world has closed in. All I am aware of is my immediate surroundings and the muzzles of their weapons.

The only sound is coming from waves lapping against the two boats. No fishing boats have left the harbour. Even on Sunday there had been boats coming and going, but today there is none. So, the four of us sit in our respective places doing nothing. We stay like this for two hours. It seems an eternity. I keep telling myself everything will be all right, it's the only thought in my head as I stare at the deck as if in a trance. A low thudding noise penetrates the world I inhabit. It is rhythmic and getting louder. It seems to be coming from behind the headland. My captors, *(because this is how I think of them)* must hear it as well because they liven themselves up. No longer lying around having a siesta. They straighten their uniforms and try to look more alert and efficient. I stare at the headland waiting to see what is coming. Soto pantomimes I should gather up my things and put them in a bag. By things he means my tablet laptop etc. Anything I consider being of value to myself.

He acts out holding the bag close and puts a finger to his lips pointing to it. I get the gist of what he's getting at. "Venezuela everyone takes, you keep safe."

Besides my devices I put all my note books and SD cards that hold the videos I made into the bag. After the disaster of the accidental deletion of videos in St Martin these are important. It is the record of my journey so far and I could not stand to lose any more. I tuck my wallet into my jeans, it only holds twenty-eight US dollars, one-hundred EC dollars, and my bank cards. I hope they will not confiscate them. They still have my passport and ship's papers. I asked for them but got no response. The throbbing noise coming from the headland is getting louder and attracts my attention.

I cannot believe my eyes. A bloody great gunboat comes into view and heads straight for us. It looks impressive with its bow wave surging up either side, but also disconcerting. As it gets closer I can make out a 20 mm cannon on the bow. It is coming up on us so fast I think it will ram Mae. When it turns and sweeps past us, the wake sends Mae and the patrol boat heaving from side to side. I thought the small patrol boat might get swamped and from the way the soldiers reacted, so did they.

They panic and shout. I can see a couple of faces looking down from the bridge of the gunboat. There is no reaction from them at all. It circles us and comes to a stop on Mae's starboard side, bow on. I have a great view of the canon pointing at Mae. It might be my imagination, but I think they are making a point. Whoever is at the helm then puts the boat into reverse and manoeuvres it alongside Mae.

A big guy with sergeant's stripes on his arm jumps onto Mae's deck. He mimes for me to get a rope that can tow Mae and then points to the anchor chain and makes a winding motion with his hand. I get him to understand that the electric windlass will not work, and I need to raise the anchor manually. The gunboat by this time has drifted right next to Mae, and she is bouncing off it. I shout and wave my arms for them to back off which they do, but not until they damaged a couple of stanchions on my port side.

I look at the big sergeant he shrugs, and signals for me to go up to the bow with him. Soto gets there before us, he grabs the anchor chain and pulls on it getting nowhere fast. I think he is trying to impress the sergeant. I signalled for him to wait until the swell eases the chain. The sergeant speaks to Soto in Spanish telling him to go back to the cockpit, which he does looking sheepish. Then the sergeant and I pull on the chain together. We get into a rhythm and after ten minutes we have the anchor on the deck. I then help him secure the towrope, not happy that he winds it around the cleats on the bow and then around the windlass. I let him know this, he takes no notice and points to the helm. Meaning for me to steer the boat.

The sergeant tells Soto, "stay with the gringo." Then goes back to the gunboat. The small patrol boat casts off. I worry the cleats and the windlass will get ripped out of the deck but there's nothing I can do. This fear is insignificant once we leave. There is no slow take-up of the slack. They speed off, and I watch the towrope snake away, then with a violent jerk Mae is moving. The gunboat picks up speed with Mae bouncing around behind it. There does not appear to be any concern that she could broadside, capsize, or swamp and sink. Whoever the skipper of the gunboat is, just wants to get where ever he is taking Mae as fast as possible.

Hanging on and trying not to get thrown about takes concentration, as Mae bucks and rears, crashing into the waves as they pull her along. Until with her rudder at amidships Mae calms down and follows like a lamb to slaughter. Now I have little to do, just make the odd minor change.

As we clear the headland, the sea becomes lumpier and Mae buries her bow more into the oncoming waves. Glancing behind me at Soto, he is no longer standing. He has wedged himself into a corner at the back of the cockpit. He doesn't look like he is enjoying the ride, the cocky look has gone. The smirk he had on his face before now a look of fear and sickness. This pleases me, so much for the big man and his gun! Towed along the coast for another hour until a bay opens on the starboard side. We carry on across its mouth.

Halfway the gun boat slows down and makes a sweeping turn into the bay.

In my mind I thank the skipper. It had been a rough start, but he obviously knows what he is doing. Now we are in the bay proper, ahead I see buildings and other structures. To the port side there is an outcrop of boulders forming a small inlet that has a sandy beach. Above the beach and too the right a couple of small buildings stand in front of a cliff face. On top a lookout tower. Hills surround the bay and straight ahead, sitting below them is the town of Juan Griego.

When Mae is two-hundred yards from the fuel dock just past the small beach, the sergeant hails me from the gunboat. Using sign language, he wants to know Mae's draft. I shout across to him dos metre. He understood and signals I should drop anchor at the same time he lets go of the towrope. A group of soldiers run along the fuel dock and are shouting to Soto onboard. He puts his arm around me and calls back. "El Bandito, Amigo" and laughs. He is the man of the moment and puffs himself up like a peacock.

The small patrol boat that pulled me over in La Guardia

comes alongside Mae. I'm told to lock up Mae and get my things. He again makes a big show of telling me to keep them close. We then board the patrol boat and go to the dock. Once ashore the three soldiers who arrested me stand next to me while the others take photographs. I am their trophy, they all want to be in a photo with El Bandito! I smile all the time but inside feel sick. From their comments. "Anglaise Bandito, mucho gangster," and excitement, they think they have caught a drug runner with a large stash hidden on the boat.

Silvio and Torres's words come back to haunt me again. "If the military get hold of you, you will not have a pleasant moment. If they cannot find drugs or contraband, they will plant some on you." They finished taking pictures Soto leads me away to their canteen and tells me to sit. A soldier who had not been outside with the others brings me a cup of water. I thank him and gulp it down. The fear of the unknown has dried my mouth out again.

Moments of fear gripped me out at sea, but this is different. Here amongst these men I am helpless. Subject to their whims and fancies. From what I have seen so far there is not much going on in the discipline department. There does not seem to be anyone in overall charge until a young soldier walks in and sits down opposite me. By his manner and reaction of the others I know he ranks over them.

In halting English, he introduces himself as second Lieutenant Cortez, acting base commander. This surprises me. Without his uniform he would look like a young boy still in his teens. Another officer walks into the canteen. The atmosphere changes and the soldiers including Cortez jump to attention. He waves his hand in dismissal they relax. He is older. I guess late thirty's. "This is Capitan Juan Perez." "He wants to inform you a medico is on his way to give you an examination." Hearing a snigger from the soldiers prompts visions of a guy with rubber gloves poking around my arsehole. Cortez must guess what I am thinking by my pale face. He assures me its routine and will be fine.

Panicking I try to explain what brought me to La

Guardia. He stops me and says. "My English is minimal, I do not comprenda." With that, he walks out with the Capitan. For about fifteen minutes I suffer an endless barrage of taunts and laughter from the soldiers in the room. I do not understand most of what they are saying but shouting "Hey gringo, your Soto's brother," is amusing to them for some reason. The medico arrives accompanied by the Capitan and Cortez. The Capitan orders the soldiers to leave the canteen.

Fearing the worst, I stand up, but the medical officer tells me to sit back down." You English?" "Yes." "This will not take long." How Old Are You?" "Sixty." He wrote this down then takes my blood pressure and temperature. He puts a stick in my mouth holds down my tongue for two seconds. "Your fine, perfect health." It is the quickest examination ever. As he packs away his things I ask him. "How is your English?" So, so, he replies. "Can you translate for me?" "I need the Capitan to understand why I am here." The medico is uncomfortable and wants to leave but agrees. He speaks to the Capitan and Cortez asking them if they will listen. The Capitan nods, and we go outside.

Soto and three other soldiers join us. Drawing a map in the dirt with my finger I try to show my boats track from St Martin to La Guardia. Explaining to the medico as I do so about the engine breaking down and the bad weather I encountered. The Capitan allows me to smoke a cigarette while the medico translates. Soto who is standing nearby points at my cigarette and draws a finger across his throat. The medico sees this and said. "He is telling you smoking will kill you." I point to Soto's assault rifle and say. "So, will that." "Yes, it will, but not as slowly and painfully as your cigarette." He translates what he says, and everyone laughs. Cortez asked the medico a question. "Why am I sailing the Caribbean in the first place?" "Because I want to sail the boat across the Atlantic back to England." "On your own?" "Yes." This impresses them. The officers leave along with the medico.

I am taken back inside and told to sit down. they bring

in a skinny frightened black guy. A soldier is holding onto him none too gently. He says. "They will give you food and then take you back to your boat." Then the soldier takes him away, God knows where? I never see him again.

A woman brings me my food. A small portion of rice with bits of chicken. Another glass of water would have been nice. While I am eating Cortez comes back and sits opposite me. "You have found a mechanic to look at my engine?" He appears impatient, shakes his head, and encourages me to eat faster. As soon as I finish, he takes me back outside, waiting are two DEA agents with a sniffer dog on a leash, the Capitan, Soto, and two other armed soldiers. I feel stupid, how naïve could I be?

There is no mechanic. They will search Mae. Once aboard I unlock both cabins, then sit in the cockpit while they tear the boat apart. When they finish looking, they let the dog off its leash. At one point they think the dog has found something in the forward cabin. Soto nearly wets himself with excitement. This after all is his bust, and he will get the credit. When the dog handler brings the dog back to the cockpit shaking his head. Soto looks disappointed and glares at me.

The agents turn their attention to the foredeck. The Capitan orders Soto to take me up onto the deck while he waits in the cockpit. Secured underneath a green tarpaulin are two dinghies. Mine and one I am taking back to the UK. A couple I met in Grenada were leaving their boat in the Caribbean. They asked me to take it for them and they will collect it from Brighton my home port. Some suggested I should have unwrapped the parcel and made sure just the dinghy is in it before I left Grenada. The trusting soul, I am. I had seen no reason to do this.

Now the maggot called doubt crawls through my brain whispering, "what if?" With my dinghy lifted revealing the parcel, its wrapped in black plastic and brown parcel tape is crisscrossing its surface. I must admit to myself it looks suspicious. Soto thought so. His eyes light up, and he points at it shouting at the agents. One of them looks towards me.

"This?" "Dinghy, the same." I said pointing at my one. He pulls out a knife and cuts away the plastic. The other agent studies me while his partner works. Sweat is running down my forehead, dripping onto the deck. I must look nervous even though I am trying hard to be nonchalant. The dog, thank God, is sitting on the deck taking no interest. When the guy finishes strips of plastic lie around the deflated dinghy. The dog wanders over it. Giving the dinghy a few desultory sniffs, it then cocks a leg and pisses on it. The agents are embarrassed and apologise for the dog's behaviour and the mess they have caused. Smiling, I say to them, "no worries." "You were just doing your job, which I know is necessary."

We return to the cockpit. They report to Capitan Perez; the search is complete. Nothing found. The boat is clean.

I am told to lock the cabins, and we return to shore. Stiffness and tension have left Perez which filters down to Cortez and their men. It is very subtle; the atmosphere relaxes. I can sit on the wall outside the canteen and smoke a cigarette. A soldier keeps an eye on me when the officers leave. Captain Perez takes off in a jeep and Cortez disappears into a small building. They have left me without saying what happens now?

As I sit smoking, I notice a subtle change in the soldier's attitude towards me Now I am a curiosity. A crazy old man who sails the sea alone. No one will speak to me. I have asked for a drink of water but am ignored. It seems I'm off-limits. After half an hour Cortez and Soto come out of the office. Soto calls through the door of the canteen, "Garcia." A stocky soldier comes out to him and they go to a covered jeep. Cortez says they are taking me to immigration to get my passport stamped. I get into the back of the Jeep with Soto, we sit on wooden boards fixed over the wheel arches opposite each other. Cortez sits in front with Garcia who will drive. No one puts on a safety belt, there isn't any. Hanging on to the rollbar is my only way to stop sliding off the board as we pull away. We go through a checkpoint at the exit and follow the road into the town. Garcia is not happy and shows it. He flicks the siren on and off if anyone is stupid enough to get in the way. Pedestrians trying to cross the road jump back onto the sidewalk, cars mount kerbs. From where I am sitting I can see the speedo. Even though it's a built-up area our speed is eighty kilometres an hour. Soto and Cortez are unconcerned. I have a small view of outside. This part of the town looks grim. Buildings are in disrepair. Render fallen off walls lies on sidewalks. Once through it we travel along a highway. He cranks the speed up to a hundred and thirty. Coming up on a queue of cars waiting to get through a police checkpoint, Garcia sets off the siren. This time leaving it on as he switches to the lane for oncoming traffic. Luckily nothing is coming the other way. Barriers are blocking the road. Police hurry to move them and wave the Jeep through.

Arriving at our destination Cortez seems at a loss who he should see. We stand around for five minutes while Cortez talks to a woman on reception. She looks nervous as she talks to him. I notice people are when they see the uniform. A tall skinny man with a bald head comes out of his office and asks what we want?

Cortez takes him to one side and I assume explains this morning's events. He comes over and says in English. "Do you have any money?" "You will have to pay for your entry stamp and ship registration." I ask him, "how much will it cost?" THEN THE FUN STARTS. He doesn't know. "Do you have Venezuelan bolivar's or American dollars?" I tell him I only have twenty-eight American dollars. "This is no good, it is at least one hundred and fifty dollars and you cannot pay with American currency." "Transfer money by wire to a bank account." "What bank account?" He thinks for a minute then says. "I don't know? The governments?" "Maybe not?" "Unless you pay I can do nothing." This guy is winding me up. No way it costs that much. "How am I meant to pay if I cannot go anywhere unless the army take me?" He shrugs "I will speak to them and they will take you to Western Union. But it is late, they are closed you will have to do it tomorrow." From the corner of my eye I can see Cortez is getting frustrated. He wants to return to the base. But I press on. "Can I change my dollars in a bank or spend them in a store?" "I need to buy a few things." He looks at me as if I'm an idiot. "No, it is illegal for citizens to have American currency." "Great, so what am I meant to do?" "My friend can help you. Wait while I get him." He disappears for two minutes and when he comes back, he has a furtive looking individual in tow. "This is my friend Fernandez. He can help you." Soto and Cortez look the guy up and down with distaste. I can't say I blame them he doesn't look like he can help himself, let alone me. He speaks English in a whiny voice. "I will give you seven thousand four hundred bolivars for your dollars." He nervously glances at Soto glowering at him. "Give me eight thousand and it's a deal." He doesn't argue, we exchange the money, and he scuttles out the door. His lack of haggling convinces me he got the best end of it. On the way back to the base I ask if we can stop at a store to buy water. Cortez apologises, we cannot, he is in a hurry to get back. He has a report to write and calls to make. He also must arrange transport for me. "Where am I going?" "You cannot stay on the command at night you must return to your boat." "Someone will pick you up in the morning and bring you back." At the base I am given a small bottle of water then the patrol boat takes me out to Mae. I will have to wait and see what tomorrow brings.

Chapter 7

Marine Commando

Day 41

As expected, even though exhausted, sleep eludes me. The uncertainty of my situation playing on my mind, the whine of outboards and Mae rolling from the wake of fishing boats passing close by did not help.

Seven a.m. the patrol boat arrives to pick me up. The two soldiers on board *(Caballero and Ortega)* have no weapons, I see this as a good sign. At least I am no longer perceived as a dangerous drug smuggler. They are friendly and give me time to get my bag. I'm not sure if I am coming back?

Ashore, Caballero takes me to the office. Cortez is sitting at his desk. "This morning I have to take you to immigration but first go to the cantina for breakfast." Caballero takes me over and tells me to sit down and he will get the breakfast. Half a dozen soldiers are in the cantina having theirs. This group was not here yesterday. Amongst them is a short stocky Sergeant with a moustache. He is loud and jovial. He says something to the others and they burst out laughing. I know it is at my expense the words loco and gringo pepper his chatter.

Caballero brings me my food and water and sits opposite me. The jolly fat Sergeant comes over to introduce himself. "My friend, my name is Vega." He has bright sparkly mischievous eyes. Garcia appears at the door and tells me it's time to go. Cortez and Ortega are in the Jeep waiting.

Again, the ride to immigration is fast. The same guy who saw us yesterday takes Cortez into his office. Garcia and I must wait outside. After five minutes Cortez with a face like thunder comes out and tells us to get back into the Jeep. We are going back to base. Over the roar of the engine I cannot hear what he says to Garcia. I probably wouldn't understand anyway, he is speaking so fast. He gets a message on his phone and gives Garcia an order. Turning around he tells

Ortega and I to hold on. Garcia puts his foot down. Ortega looks excited and mouths the word. "Bandidos."

When we get to the town Garcia slows down and we cruise around the streets. Cortez spots a motorcycle carrying a passenger who has a backpack. When Garcia turns on the siren, the bike in front speeds up and turns down a side road. We soon catch up and Cortez sets off another siren. Whoop! Whoop! The passenger turns around then taps his mate on the shoulder and the bike pulls over.

Garcia and Cortez jump out of the Jeep. Cortez has his pistol drawn and the two bikers put up their hands. Ortega tells me to stay put, I see him thumb the safety off on his rifle. Then he gets out keeping it pointed at the ground, but I do not doubt he could bring it up fast. Garcia turns the passenger around and pushes his face against the wall kicking the man's legs apart as he does so. He takes the guys backpack and throws it to Ortega then pats the guy down. Cortez covers the driver while Ortega searches the pack.

Then it's the drivers turn. Garcia is rougher on him maybe because he didn't stop straightaway. Satisfied neither is armed he then holds out his hand and says. "Papers." As he reads them he questions the two men. There are no niceties. The three soldiers have no worries about infringing any rights these guys have. From what I can make out Garcia doesn't think they have any.

The two men do not back chat or look surly, they look frightened. Cortez lets them go. Garcia protests. They have a backpack and match the description given of two men on a motorcycle, the passenger sticking a gun in people's faces and demanding money. Garcia is convinced the passenger dumped the gun when they turned into the side street and argues. Cortez is not listening and orders him and Ortega to get back into the Jeep and drive back to base. Cortez apologises for this happening while I am with them but there had been no time to drop me off first. I tell him "no worries mate it's not every day I get driven through a town at high speed chasing gunmen by Aryton Senna," gesturing with my thumb up to Garcia. Cortez gets the gist of what I'm saying and translates for the other two. This breaks Garcia's bad mood and he laughs saying, "You Ok anglaise ok." I feel winning these guys over will be important for me.

It's not long before Garcia must slow down, there is a disturbance ahead. A crowd of people are spilling out onto the road from an entrance to a compound. There is shouting and screaming. A woman runs up to the Jeep asking for help. Again, I am told to stay in the vehicle, the three of them get out Ortega guards the Jeep, the other two disappear into the crowd with the woman. They soon come back. Whatever it was they had sorted it, and the crowd disperse. They did not offer an explanation, and I did not ask. Whatever it was must not have been serious they are soon laughing and joking with me on the way back to the base.

Once there, Cortez takes me into the office. Maybe now I will find out what happened at immigration. "The civilian administration says they must inspect the engine on your boat." "Why?" He shrugs. "You are to remain on the commando during the day and stay on your boat at night." "You are the commands responsibility until the inspection is carried out." "Am I under arrest?" "No, but until your passport has an entry stamp you are?" He pauses searching for the right word in English. He looks embarrassed. "Restricted?". "Do you need anything?" "I would like access to the Internet, so I can let my partner *(I tell him the wife)* know where I am and that I am well. Also, I would like to go to town and buy supplies for the boat.

He says he will do what he can tomorrow but now we will go to the cantina as a meal will be ready soon. On the walk to the canteen he points out where I can go. "It will not be necessary for you to have an escort in these areas." I take this as an improvement on my situation. Admittedly it is like being under house arrest but at least I will not always have a babysitter with me while on the base.

In the canteen I am invited to sit with him and Garcia who fusses around me. He gets me a glass of water and wants to know my story. Cortez translates as best he can. After the meal we go outside.

Other soldiers join us they listen while Garcia relates the events of our drive. Garcia is popular with the men, and they all respect Cortez as young as he is. When they leave no one calls me gringo, now its Amigo or Anglaise.

Caballero comes to get me at five and takes me back to Mae, someone will pick me up in the morning. But he cannot tell me what time. Back on Mae I tried to make sense of this inspection. Why do they not believe the engine broke down? I'm also disappointed that I still could not speak to Mid.

Day 42

A long-range fishing boat came in last night and anchored fifteen metres off Mae's starboard side. It must have been out at sea for some time. The crew were noisy and clamouring to go ashore, as soon as they had offloaded their catch. Two dinghies came from the large boat moored next to the slipway to pick them up. An old boy I assumed to be the skipper stayed on board

I fell asleep but the sound of an outboard and men shouting woke me up at two a.m. The crew were drunk, and two women were with them. The sound of Grunts, groans and the odd squeal from the women soon carried across the water. I made tea and sat on deck the old fella I saw earlier was on deck having a smoke. When he saw me, he raised his hand, and I acknowledged him. When the grunting and groaning stopped, he went below, I did the same and fell back to sleep.

Seven thirty a.m. I am impatient to get ashore the prospect of talking to Mid makes it worse. It must worry her not having heard from me since Sunday. The patrol boat arrives at eight. I go straight to the office, but Cortez is not there only Morales and Soto. Who I now realise is pissed off. It has put his nose out of joint; he is no longer the hero of the hour who arrested a drug dealer. When I ask if I can use the Internet he gets snotty and says no. Gonzalez looks up at him and frowns then tells me to go to the cantina for breakfast.

The jolly fat Sergeant called Vega is there, sitting with Garcia. Vega says, "you want coffee, come." Leading me into the kitchen he introduces me to the woman who cooks for the base *(Leah)* as my friend from Londres. Then tells her to bring breakfast and coffee for three. He has obviously been talking to Garcia about me. Vega has limited English but somehow, we communicate. He wants to know what I have in my bag?

I show him the tablet. His eyes light up when I turn it on and he sees the Google logo. He sees the answer to our language barrier "Googly, googly translate." Garcia asks him a question. When he realises what Vega is on about he is eager to start it up. I get across to them without the Internet it will not work, and Soto said I can't have access.

Vega's lip curls, and he raises a hand. "Vega, master sergeant." He lowers his hand. "Soto, private. Wait my friend." Garcia grins as Vega marches out of the canteen. I feel this will cause a bigger problem than I already have with Soto. Two minutes later his back with the Wi-Fi code. We pass the tablet back and forth and I learn a lot about them, even though the translation is not great. *(The technology is still new)*.

Garcia is in his early thirties and married with one child, a little girl. Proudly he shows me a photograph. Vega is forty-two married with two children a boy and a girl. The base commander Cortez is only twenty-one. I'm quite shocked by this. Such a young man in charge. Vega is a master sergeant, Garcia is a corporal and Cortez's driver. Caballero is a sergeant along with Morales. Ortega and Soto are Privates. Vega says Soto has enough ambition for everyone, be careful of him.

They warn me not to take photographs of the base or any soldiers, they do not permit this and would cause me a lot of grief. Vega wants to know why I have problems with immigration and what's wrong with my boat? After I explain it to him his eyes sparkle, and he tells me he can help me.

I know this guy is a rogue and will not be doing something for nothing, but I cannot help liking him. Garcia says with a serious look on his face. "Vega is a man who knows low people in high places." Then laughs. "I need to contact my wife." "How you going to do that?" asks Vega. "Skype." The UK is five hours behind, it's only three thirty in the morning there, I don't know when I will get another chance. Mids sleepy voice answers. Vega and Garcia leave the cantina to give me privacy. The connection isn't great I tell her what has happened. It alarms her, but she is relieved I am all right. We only talk for a few minutes before I must hang up. Soldiers have arrived, I tell her we will talk later.

There is nothing comfortable to sit on only wooden slatted chairs or hard plastic ones. The wall under the tree outside the cantina is the best place for me. There, Wi-Fi reception is better, and I can observe the routine of the base without getting in the way. A problem is recharging the tablet. The power supply here is worse than St Martin.

There is nothing to do. I wish I had put a book from the boat in my bag, I must try to remember to do it tomorrow. Three hours pass then Vega comes to take me to the office, there is a friend of his waiting to see me. Cortez is also there. The guy's name is Josef, and he speaks fluent English. "Vega has arranged for a mechanic to look at your engine." "He can also take you to someone who can help you with your money problem." "Thank him for me, I will appreciate it." Vega's smile beams out when he hears my answer, but Cortez has a frown on his face and speaks to him. Joseph translates. "The commander reminds Vega that the mechanic cannot touch your engine until after the inspection." "I understand of a night they put you back onto your boat?" "Yeah that's right." "They wonder if you get lonely and would like a woman?" "No thanks I'm fine." "You dislike Venezuelan women?" "Venezuelan women are beautiful, but I imagine, *(rubbing my thumb and fore finger together)*, are expensive?". "Not at all, you can have one for free." "No thanks, I'm okay." Josef raises his eyebrows and says. "Maybe you would like a boy?" He sees his question has caused offence as my tone changes. "No, make them understand I have a wife who I love very much." "She is also my best friend." "I do not betray my friends." This hits a nerve. Their sense of honour. I get a look of approval from Cortez and the others when they hear Josef's translation. Cortez asks him to tell me he will take me to town late this afternoon. Vega will arrange for you to see his friend about the money. Josef shakes my hand and leaves. Going outside, Garcia follows. He asks me if I will roll him a cigarette? Rolling two I pass one to him then sit on the wall and try to call Mid but cannot get connected. The Wi-Fi is down. Vega joins me. He tries hard to teach me some Spanish and gets frustrated. Feeling the same with his effort of learning English we agree to give up for now. A civilian approaches us. Vega nudges me, nods towards the guy and hisses under his breath, "Gimenez." Then stands up. "Hola Gimenez." "Buena's Tardes." They are both acting, I can see there is no love lost between them. It's all for show. Feeling like prey while two predators circle around me I say. "Hola." "Ah, the Anglaise my colleague told me about." Now he has switched to English Vega watches him like a hawk. "Your colleague?" "The Capitan you see at immigration." "Oh yes, I know him." *(I get the impression Gimenez is as slimy as his colleague).* "You will see me regarding the inspection." "Do you know when it will be?" "Who can say these things take time." He smiles and says goodbye in English to me and Spanish to Vega. Pretence over. Vega glares at him as he walks away and snarls. "Civilian Administration." Then spits in the dirt.

It's four p.m. before I am taken into town. We stop outside a supermarket, Cortez comes in with me. It's a shock to see a supermarket with only half its shelves filled. There is not much choice of what there is.

I pick up a case of water and bags of sweets *(crappy ones)* packet of cakes two bottles of lemonade and three packets of cigarette papers. There is no milk fresh or dried. No tea, coffee, or bread. Silvio had told me about the shortages, I didn't realise just how bad it is. Cortez says.

"There is not much so people go without." "The only thing we have plenty of is oil and gas." Meals on the base are frugal. The best has been a small amount of rice with a few pieces of chicken. When we get back, there is no time to contact Mid. I must return to Mae

Day 43

Today is not going well. The up and down roller coaster ride of emotions while waiting for a boat to take me ashore is mind numbing. No one comes until ten.

Cortez allows me to use the base internet. For the first time since leaving Grenada I can connect with a video call and see Mid on Skype. All previous contact has been with the landline or mobile. keeping myself in check seeing her after all this time I try not to get emotional. She says, "Yesterday, I spoke to the British Foreign Office and informed them of your situation." "They will contact the embassy in Caracas who will help." *(Not that I think the British authorities knowing where I am can do much, but it makes me feel better).* "What are you doing about money?" "Can you use a cash machine there, or shall I send some?" Trying to explain the way money works here is confusing. "What do you mean?" "I don't have time to explain now, but sending money is not a straightforward process." "Within the next few days I hope to have a secure bank account you can send money to. I will let you know when?"

Vega and Caballero arrive and stand behind me. Assuming I am looking at a picture. Caballero says, "your wife?" "Yes." "Say hello Mid." Impressed when she waves and says hi. Vega says, "hello pretty lady greetings from Venezuela. We are friends and will look after Roy." "Thanks, please do that." Ortega and Soto come out of the canteen and join us. I know Mid gets a shock by the sight of the men in uniforms holding assault rifles standing behind me. But she does not show it. They all wave goodbye and leave. "Are you sure you're all right?" "Yes, fine, don't worry." By the look on her face I know she is. "I have to go now, if possible I will call you tomorrow." Giving a final wave, I end the call.

I know it surprised them to see Mid live on the screen. They did not know the tablet could make video calls. The technology is here but I suspect many cannot afford it. Everyone here appears to have basic mobile phones.

Still having the Wi-Fi connection, I call Oscar in Grenada. "Where are you?" "Expected you back weeks ago." "I worried about you, that storm was terrible." "On a military base on Margarita island Venezuela." He doesn't ask why but says. "Oh, shit!" "When will you return to Grenada?" "How long is a piece of string?" "Even when I know it may not be possible to tell you." "The base internet is unreliable, I never know when I can get connected". "Anything I can do?" "Not that I know of, I wanted to let you know I survived the storm and am alive." "Mid has let the British Embassy know I am here, maybe they can help?" He tells me to take care then I lose the connection.

Because it was late when they picked me up I missed breakfast. Hungry I wander into the cantina to see Leah. She has finished her shift and is about to leave but before she does, she gives me a glass of water and a cold arapa. Which is a flat bread made from corn flour? She makes them fresh every morning, they taste better hot. She apologises there is nothing else. Going back outside I sit on the wall, munch on the bread, and read my book. *(I remembered to put one in my bag last night.)*

Morales who is coming out of the office sees me and comes over. Would you like a shower? "Would I?" "Yes please." The last one with fresh water was in St Martin eighteen days ago. Holding a finger to his lips he leads me through a door next to the canteen, it's a dormitory half a dozen beds line both sides of the room. A locker is beside each one. Half occupied by sleeping soldiers. At the back of the building are the showers there are two, basic is an understatement. There is a small encrusted showerhead attached to a single pipe that connects to an old brass tap. There is no evidence of hot water. He gets a towel from his locker and a sliver of soap. When I turn it on there is not much pressure, the water dribbles from the shower-head. It doesn't matter freshwater on my skin feels wonderful. When I am finished, I give Morales back his towel and soap thanking him.

He takes me back outside and mimes cutting my hair. I pass on that, even though I need one, I don't think the Venezuelan military haircut would suit me. So as not to offend him I tell him I like my hair as it is.

For the rest of the day I read my book, there is nothing else to do. Before I am taken back to Mae Vega tells me, we will go to see his friend who can help me tomorrow.

Day 44

Eight a.m.; A fishing boat pulls alongside; the man doesn't speak but points to the dock. "You are here to take me to the base?" "Si." Looking towards the dock Vega is waving. Getting my bag then locking the cabin I jump aboard. This guy doesn't seem happy about picking me up.

When I get ashore Vega encourages me to hurry. Cortez, Soto, and Garcia are in the jeep waiting. Hola's all round, then we're off. "We go to see my friend, ok" says Vega. Cortez looks apprehensive and I wonder what's wrong. Soto is trying to be mean and moody, but I can see he's excited. Vega is Vega, beaming his smile and talking Garcia's ears off. Just past the checkpoint, Garcia turns left onto a road leading away from town. Soto seeing my questioning look says. "No Juan Griego, Santa Anna." Pulling up outside a scrapyard Vega and I go in while the others wait in the vehicle. Vega introduces me to a well-built guy who speaks English. He tells Vega to come back in an hour then he will take us to see his father. At the Jeep Cortez keeps looking at his watch and fretting about how long this will take. Vega says stop worrying it will be ok. After an hour the guy comes to the Jeep, Garcia stays in the vehicle the rest of us get out and walk with him to a store.

Cortez is nervous, I can understand why. None of what

will happen is legal. He wants nothing to do with this. The big fella takes us to an office at the back of the store. *Walking in is like entering the set of the Godfather and I am meeting Michael Corleone.* The father is as big as his son and is sitting behind a massive carved mahogany desk.

His hair is black, touches of grey at the sides and swept back from the forehead. Smoking a large Cuban cigar, looking the part. Vega rushes over to him and shakes his hand speaking so fast I cannot grasp any of what he says. Cortez doesn't come in, he waits just outside the door. The son invites us to sit on the couch opposite. He sits on the desk to the right of his father. Soto doesn't seem as full of himself in their presence. The barrel of his rifle is pointing at the son who leans across and pushes it down to the floor while wagging his finger at Soto. Soto goes pale and can't stop apologising.

A woman brings in fresh coffee and serves it. Once she leaves, the son says. "My father understands English better than he can speak it so, I will answer for him." "Vega has explained your problem and how you come to be here." "But he would like to hear it from you." He listens while I tell my story. When I am finished, he speaks to his son who then says. "My father says he will help you, what is the amount you need exchanged." "Four hundred pounds." His father taps on a calculator then shows it to his son. He looks surprised. "That is only about six hundred dollars. Are you sure it is enough?" "It is all I can afford. This trip has been expensive."

The father writes on a pad tears the page off and hands it to his son who reads it then passes it over. "Here is the account number and code you will need to transfer the funds." "Be sure you get none of them wrong or you will lose your money." "You are not familiar with how things work in Venezuela?" "No, but I am learning." He laughs. "Yes, things must seem chaotic to a foreigner." "Your money will pass through Panama, Uruguay then Miami, it is the only way." "We are doing this as a favour to Vega, but also because my father likes you." "Maybe one day if we come to London you will do a favour for us?" I smile hoping it doesn't look too sickly and say "Certainly." "Don't worry Roy, I joke with you." We all stand up and shake hands except for Soto, they ignore him.

We go back to the Jeep, but Vega stays behind talking. Cortez is impatient and tells Garcia to sound the horn. A minute later he comes out of the store and hurries to get in. On the way back to the base Soto keeps asking Vega questions until Cortez turns around and tells him to shut up. It's easy to see Cortez isn't happy.

We get back in time for lunch. Garcia and I go to the canteen. Cortez and Vega to the office with Soto following.

After five I get through to Mid and give her the bank account and transfer numbers, emphasising how important it is not to get any wrong. As the UK is five hours behind the banks are open, she says she will deal with it today. Soon I will have to go back to Mae, I must wait until tomorrow to find out if the transfer goes through okay.

Back on Mae I wonder what this favour will cost Vega.

Day 45

This morning I am impatient to get ashore. As soon as I can I will contact Mid, I want to know how she got on at the bank. At eight o'clock I see four unfamiliar soldiers all armed get into the patrol boat. For one horrible moment I think something is wrong and they are coming to get me. But instead they head out the bay. While I watch them disappear in the distance I did not notice two old guys in a fishing boat had pulled up alongside. "Signor, we take." Says one of them pointing to the dock.

Ashore I go straight to the canteen for breakfast. Then sit on the wall outside killing time until I can call Mid. Vega is getting on my nerves, every half hour he pesters me to call her. He wants to know if she has sent the money. I can't get it through his head the UK is five hours behind and she will be asleep.

Midday I call and It's not good news. Mid sounds

agitated "I gave the numbers to the cashier stressing how important this transfer is. She told me it would take five days to complete." I interrupt. "Five days? That's a joke." Mid carries on, "When I got home I checked the receipt, the first number of the transfer code is wrong." "Panicking I rushed back to the bank but was too late the money had gone." "After an argument the manager came to see me." "He did correct the mistake but now it could take thirty days to get through." My hope of leaving here the end of next week dwindled away leaving me feeling numb. "I'm sorry Roy." "It's not your fault." "What are you going to do?" "I don't know?" "have you seen anyone from the British Consulate?" "No, not yet." We talk for a few minutes more, then I go to the office to tell Vega the news.

Before I can say anything, he asks. "Have they sent it?" "Eh, yes and no." Using Google Translate I explain what is wrong. "But they sent it?" "Yes." Flopping into a chair I stare at the floor, my mind blank. A slight grey-haired man enters the office. Looking at no one in particular He speaks in Spanish then says my name. Ignoring him I carry on looking at the floor, assuming he is just another civilian official come to make more trouble for me. Then I hear the words British Consulate. Vega points at me. "Roy Cleeter?" I nod. "My name is Harry Lees. I am the British Consulate." Standing up I shake his hand and cannot resist hugging him. Thinking, finally! Someone will do something to get me home. The embassy in Caracas asked him to find out who is holding me and my circumstances. He asks if I have any money to pay immigration? "No." "How much did they want?" "One hundred and fifty US dollars." He raises his eyebrows. "That doesn't sound right, who told you this?" "A guy at immigration in Pampatar."

"British Embassies no longer have the funds to help citizens in distress in foreign countries, so I cannot pay it for you." "I am working from home, contact with the Embassy in Caracas is via the Internet or mobile phone." "Grenadian and Antiguan fishermen who wash up here or the odd tourist who gets robbed are who I deal with." "But this is a rare occurrence as British tourists or any others have stopped coming." "I can only help to get money sent from their families and act as an intermediary between them and the Venezuelan government." "So, unless you have someone in England who can send money there is not a lot I can do." "Do you have a bank card?" "Yes." When Cortez arrives, he asks if he can take me to a bank with an ATM to see if I can get any money. He agrees if one of his soldiers accompanies me. At the moment none are available. Harry says he will wait until one is. In the meantime, will it be ok if we wait outside?

Harry makes sure we are away from the windows and not in earshot of anyone who may be listening. "Tell me how you come to be here and what has happened since they picked you up?" He listens while I give him an outline. "My advice is trust no one." "You are in a dangerous situation, nothing is as it seems." "People will claim to be your friend and want to help you." "In rare cases this may be true, but when you are dealing with officials or the military forget it." "They all want money or whatever they can get from you." "Just be careful."

Harry falls silent when Cortez interrupts us. An off-duty soldier is with him who will escort me to town. Harry's car surprises me, I did not expect a British Consulate to be driving around in a beaten-up Mondeo. He lifts the bonnet and reconnects something. When we get into the car, he says "I have to do that it's my anti-theft device."

At the bank, a long queue is at the ATM. Harry tells me. "This is normal, let us hope the machine does not run out of money by the time you get to use it." When my turn comes Harry shows me how it works. "Do not press the button for the English translation." "If you do, it will shut off." "When you have put in your pin code, it will ask you to put the first or last two digits of your passport number." "Make sure they are right, or it may keep your card." "They limit the ATMs here to giving you six hundred bolivars." "You may get lucky if you present your card again it might give you another six hundred." I wasn't, the second time I try the machine spits out my card and closes. Harry asks the soldier if it is ok to go into the store next to the bank? He agrees, I need to buy more water for when I am on the boat. We get back to the base and Harry leaves, telling me. "Try not to worry I will be back tomorrow." "How Old Are You?" "Wednesday I will be sixty-one." "Oh, happy birthday for Wednesday." Giving me a shrewd look, he says "I suppose this trip is your swan song, your last big adventure?" "Well, let's hope not your last?" "Thanks, your right, although I was just expecting to sail the Atlantic and all the thrill that entails, not this?" "Seriously you should write a book, when you get back." *(I noticed he was careful to say when, not if.)* "I would like a copy". "Just don't publish it this year please, it might cause a diplomatic incident." We both laugh, and I promise him I wouldn't. With Harry gone Cortez asks if I want to go back to Mae? After the bull shit I have suffered today I am glad to.

Sunday 28th Day 46

I am so fed up; my novelty value must be wearing off. It's ten a.m. and no-one picks me up. I am torn between the desire to leave Mae and fly home or stick it out here for as long as it takes and sail back to Grenada. The airport on Grenada is only a few hundred miles away, where I can get on a plane that will take me to Gatwick, But, it is across open water. What waits for me out there? Another storm? Pirates? Questioning myself, has my confidence gone? Will I be apprehensive? No, let's say it like it is, SCARED! Thinking about it, I feel queasy in the pit of my stomach. This cannot end in disaster. Because of the screw up by the bank I cannot leave here until the end of July. Even that estimate could be optimistic. Having to wait until the money from the UK arrives and the civilian administration to decide what I can or cannot do. I have got to stop thinking about it. It's too depressing.

Looking across to the base it looks deserted. No vehicles parked and the only soldier I can see is the guard outside the office. Three p.m. I get out the barbecue and cook macaroni with an oxo cube. The result doesn't look appetising, I try to imagine its Mids shepherd's pie with lots of greens and a thick rich gravy followed by a trifle. but it doesn't work. It is what it is, beef flavoured glue. As a substitute for the trifle I use an old *(as old as me by the look of it)* sweet potato and honey, it will have to do. After my meal I look around the bay. It is beautiful if you ignore the military base. Pelicans are diving for fish, Barca's *(Small fishing boats)*, lay up on the beach. Around me three bigger long-range fishing boats are at anchor. There are no cruising yachts here except for Mae and one other, anchored off my stern, stripped of most of its gear and now a pelican roost. It is not an old boat, looks like a Bavaria, smothered in pelican shit. How long it has been here and the fate of the owner I can only guess. I have asked the soldiers, they just laugh and shrug. Maybe this will be Mae's fate. The temperature never drops below twenty-six centigrade of a night and during the day the average is thirty-six. If things were right in Venezuela people would pay lots of money to have a holiday here. But things are not right, and I want to get the fuck away.

Day 47

Picked up earlier than usual this morning and taken to Cortez, He apologises for yesterday. There was a situation he had to deal with and could spare no one. "Will your consulate come today?" "No idea?" I ask, "Has anyone notified you if the inspection will happen today?" "No, maybe I will hear something later?" "Go for breakfast in the cantina."

By mid-afternoon I know nothing will happen. Losing the internet connection while talking to Mid, I sit on the wall cursing the Wi-Fi. The Consulate shows up and asks if I am all right? "Not really." "Did you see that official?" Sighing and looking as pissed off as I feel. Harry says, "Because I am the British Consulate here does not mean I am high on any list for people to see." "The official was not available"

"I will try again tomorrow."

He sits down. "Let me explain." "Maritime law dictates any seaman washed up on a foreign shore through no fault of his own. Should get aid to enable him to leave." "If not by repairing the vessel, at least to leave by other means." "This is not international law but a moral obligation that many countries adhere to." "If you were Cuban, you would have had a full medical, put up in a five-star hotel and fed the best food and a flight home." He gives a rueful smile. "Probably on the presidential plane." "Think yourself lucky you're not an American, or you would be in jail or could be face down in a ditch with a hole in the back of your head." "You could not have arrived here at a worst time." "The Venezuelan government don't like America and Great Britain too much at the moment." "They expelled two diplomats from the embassy in Caracas a few days ago." "When your problems get sorted out, I advise you not to leave the boat here." "There is nowhere safe to leave it." "All the marinas are prone to having the boats robbed and or vandalised, also I advise, leave as soon as possible and put as much distance as you can between yourself and Venezuela." "The longer you stay the more your personal safety is at risk." "You know I cannot get any repairs done until after the inspection?" I have doubts about getting any parts I might need for the engine. Harry says, "If I am still getting the run around after tomorrow I will get in touch with the embassy in Caracas and see if they can apply any pressure." I raise my eyebrows. "Yes, yes, I know, I wouldn't hold my breath either."

After he leaves, Vega wants to know what we talked about, he seems on edge. Because the Wi-Fi is down, we can't use Google Translate so speaking is difficult. He tells me he is taking me into town. A friend of his who has a taxi is coming for us. While I sit in the back Vega speaks, I know he is asking questions, but he is talking too quick, I don't understand? We do not go to town his friend turns off the main road and we head away from it. After fifteen minutes he pulls up outside a small house. A woman is sitting on the veranda. Vega gets out and speaks to her then waves us over. Her name is Kate, she is from Trinidad and speaks English. Vega has asked her to translate for him. It becomes clear what is bothering him. She says, "It worries him you told the consulate about the money exchange he arranged." "If he tells the authorities Vega will be in a lot of trouble." Looking at him I can see it concerns him. "Vega my friend you need not to worry." "He knows I am waiting for money to come from England but no details." Besides his only interest is in getting me home. It takes a while to convince him, but eventually I do. We stay at Kate's for coffee and it gives me a chance to get to know her and the taxi driver. His name is Ernesto.

Arriving back at the base Morales is waiting to take me to Gimenez, he had telephoned the base and asked that I come across to his office. When I get there he is waiting outside, all smiles and friendly. He says with a smug look on his face that under no circumstances will the inspection go ahead until I pay the immigration fees, he mumbles something about, "gusto". I asked him "what is gusto?" He shrugged and walked back into his office. I know damn well what gusto means. *(A bribe)* He has no chance, slimy bastard.

Morales brings me back, and I go to the cantina. Leah gives me a meal. She sees how pissed off I am. She asks where I have been. "Gimenez's office." She raises her eyebrows and mutters obscenities under her breath. Leah is no fan of the civilian authorities. I have yet to meet anyone who is. The meal is Arapa, with a fried egg on top. Slipping another egg on my plate and giving a wink she holds a finger to her lips. Leah could get into trouble by doing this, she apologises there is no coffee. There will be no more for a while unless one of the sergeants can get some on the black market. When I finish my meal, I go outside to sit on the wall and wait for someone to take me back to Mae for the night.

Day 48

I wake up feeling depressed, miserable, and homesick. Missing Mid and my family.

Eight a.m. Vega meets me with a grin. "Googly translate". I give him the tablet and wait while he writes what he wants to say. *"I organise transport to take you into town."* He will come along with Zapata, and a driver, so I can go to the bank. At the bank I join a long queue waiting to use the ATM. When I get to use it, the damn thing has run out of money and shut down. Word soon filters down the queue and people rush off to the next bank.

Vega sees my frustration and suggests going to the store come cafe just down the street. We have a drink and I see if there are cigarette papers for sale. There is, but more expensive than a packet of cigarettes. Only having two thousand Bolivars left in cash and papers costing four-hundred and eighty I dither. But I have no choice, having a smoke is the one thing keeping me sane. Vega surprises me by buying a packet of cigarettes and handing them over. Finishing our drinks, we go back to the car, I thought we were heading back to the base. But Vega has other ideas. giving instructions to the driver, five minutes later we stop in a side street. Telling me to get out and follow him, ordering Zapata and the driver to pull around the corner and stay in the car.

Walking down the road, I tag along beside him with no idea where we are going or why. As soon as the car is out of sight, he doubles back and stops outside a double fronted shop. "You need money yes?" I nod. "My friend can help." Walking in Vega speaks to the girl behind the counter. She takes us to the back of the shop. There is a young man and an old guy sitting at a large coffee table, they are father and son. In the middle of the table is a hookah and the old guy is using it, blowing out clouds of smoke. Vega introduces me and carries on speaking. The old guy just sits smoking and listening, the son asks Vega a question then looks up at me with interest. I assume Vega is telling my story. the old man looks at me and says in perfect English. "My name is

Hakeem, and this is my son Ahmed, nice to meet you." How much money do you need? I was not expecting this and needed time to think. Having no idea how much is in my bank or what the exchange rate will be. Hakeem mistakes my indecision for a reluctance to do the deal, he says in an offended tone; "Do not worry, my son and I will not cheat you, Vega is our friend and vouches for you." I assure him I am not thinking it at all. I explain I am just unsure how much I can afford. Ah, I understand, money is always a difficult problem for everyone. I take a chance; "Will you be able to let me have thirty thousand?" "You can have double if your bank says yes." He asks if I have a credit or debit card, and my passport with me. "Yes." "Thirty thousand, there will be a charge of three thousand is this okay?" I agree.

His daughter takes my card and passport. Taps the card number into her machine then my passport number. My heart is in my mouth waiting for it to respond. *(Is this an elaborate scam? Is my card and passport being cloned?)* When it did, she gives me the receipt to sign. Then gives me back my card and passport. Hakeem shouts out to his son who disappears through a door at the back of the shop. Five minutes later he comes back with a small cardboard box which he hands over. I look inside. It contains bundles of notes. After the handshakes and thanks all around, we go back to the Jeep and head back to the base.

Once there, we go into the canteen. Vega insists we count the money to make sure it is correct. I am uneasy doing this, anyone could walk in. Thirty thousand bolivar's is a lot here when you consider that seven thousand a month is the average wage. *(Approximately fifteen pounds Sterling)*. Splitting the money and wrapping it into three bundles, I put it back in my bag. I now may have enough money to pay immigration and a mechanic to fix the engine. Leaving me some to live on. *(I hope)*. Until the money transfer is completed. Vega says he will take me to the immigration office in Pampatar tomorrow. Eager to get back to the boat and hide the money I ask him if he can arrange for me to go now. He goes to the office to find out. When he comes back, he gives me a slip of paper. It's a message from the consulate saying, *(he doubts if he can make it tomorrow, he must go to the prison and see a fisherman from Grenada who is being held there. If he can't, he will come Thursday to see how I got on).* I would be happier to have him with me. There would be less chance of me being ripped off again by that slimy bastard and his mate who seem to be the cause of the shit I am in. Vega says, the patrol boat is out but a young fisherman will take me in his dinghy. On the way, two of his mates pass us in their boat and asks, "Why are you taking the Bandido back to his boat?" He points to Vega standing on the fuel dock watching and says. "Commando." They pull away in a hurry. When we get to Mae, I give him one hundred Bolivars. Onboard I hide the money and settled down for the night

Chapter 8

Happy Birthday

Day 49

Today is my sixty-first birthday, I wake up at six thirty and get ready to go ashore. I put on socks, boots, and jeans for the first time since leaving St Martin. It feels strange and I don't like it, but I suppose I should get civilised again. It appears the soldiers on the base are getting ready for some event. Two marquees' have been erected and a podium with big cabinet speakers in front. I am not holding out a lot of hope of being picked up in time to get to the immigration office in Pampatar. Maybe it would be better if I am not taken today? It will probably be a waste of time anyway.

I would like to go into Juan Griego though and see if I can get my phone fixed or buy a new one now I have money. Then I could contact Harry without having to be on the base and relying on the dodgy internet connection. Sometimes I am not given the correct access code, whether this is by accident or design I do not know but suspect the latter. A new phone will cost about six thousand Bolivars, that's about thirteen pounds. It will be useless once I leave Venezuela but would save me a lot of aggravation while I'm here. Vega told me to buy a phone here is easier said than done. Why though?

Nine o'clock and still no sign of anyone coming, the activity on the base has increased. More soldiers have arrived and there is more stationed on the checkpoint just before the entrance. People are arriving, the guards making them join a queue which stretches up to the top of the hill overlooking the base. Another group wearing bright orange sashes are herding people like sheep. At first, I thought they were civilians acting as marshals, but appear too rigid in their formation. I think they are a para-military group that operate here, brought over to the base to help the army.

I am recording what I see on the tablet but only on audio.

To film it would be too risky. I cannot afford to have it confiscated. It is my only means of contacting anyone, and I don't want to get accused of spying. At nine o'clock one of the bigger long-range fishing boats appears at the entrance to the bay. The small patrol boat with Cortez and half a dozen marines on board race away from the fuel dock to intercept it. Cortez orders the crew to heave to and with four of the marine's board the boat. I am too far away to see much happening, but the marines have unslung their rifles and are covering the crew while Cortez goes below with the skipper of the boat. Several minutes later the boat can carry on and the patrol boat returns to the dock.

Trucks are now arriving on the base loaded with what looks like bags of flour and other foodstuffs. Civilian officials sit at the tables set up beneath the marquees. the people arrived at seven this morning. Not until 10 o'clock do the soldiers at the checkpoint let in several people. They go over to the officials, sign a paper and then collect the commodities they allocate them. Not until the last person has gone back through the checkpoint, do the soldiers allow another group through. How have they stood waiting in this baking heat for so long? It suggests they must be desperate. Music is being played over the public-address system. Interspersed with a guy making announcements, slagging off other countries, America being the prime one. He then praises the president saying what a good and generous man he is. Giving free food to the population. The songs get more patriotic, and the commentator is giving it to the Yanks. What the fuck has happened? If I had a phone, I could ring the Consulate and get information. I hope this does not involve the UK. People keep arriving and trucks bring more goods for distribution. The food so needed by the people soon runs out. By twelve thirty it is all over. A lot gets nothing. One thirty Corporal Morales arrived on board a fishing boat to take me to the base. Once there, he says, "Go to the canteen, you are too late for breakfast, hurry you may be in time for lunch." Lunch consists of a small portion of spaghetti and two small pieces of chicken. There is no coffee only water. When I finish my meal, I join Vega and Pena *(The colour sergeant)* who are sitting on the wall outside. Vega has a Spanish to English phrase book and wants me to learn, I ask them about this morning. They do not want to talk about it and say nothing, so I do not push it. Everyone on the base seems tense and preoccupied today.

Cortez who is normally friendly comes over and says. "The Consulate can be at the immigration office tomorrow to help get your passport stamped." "Whether it will be possible to take you, is another matter." "I am sorry." "Go back to your boat now." He then walks off, I go to ask him why? But Vega nudges me and shakes his head, so I stay silent. Vega gets Morales to take me back to Mae.

Day 50

The harbourmaster of the port of Juan Griego, wants to see me this morning. Gimenez, who is the short, balding, stocky man who I met before. When I get to his office, he shakes my hand and invites me to sit down. He has a permanent smile and hides his eyes behind dark glasses. Coming across as friendly and willing to help, is an illusion. He knows I only understand a little Spanish and says, "do not worry I speak superb English." When I ask, "What is happening about the inspection and can we resolve the problem of my entry stamp?"

He answers in English with the phrase, "for example," then reverts to rapid Spanish so there's not a chance of understanding a word. Then he looks at me with a self-satisfied expression on his face which turns to one of annoyance when I reply, "Non comprenda?" Giving a big sigh he repeats the process. Starting again with, "for example." After half an hour of this, I get to understand inspection of the boat may happen tomorrow. Then again, it may not. The inspection might happen next week. But then again, it might not. Before anything can happen though, I will have to see the boss who is the director of all the port authorities in the area. I ask, "When?" "Tomorrow at ten o`clock, you see the director." "Where?" Smiling he says. "At his office of course," *(The urge to jump over the desk and strangle him I resist.)* "Where is that?" *(He fiddles with some papers and doesn't answer.)* Explaining to him I am under the control of the army as I am considered an illegal immigrant until someone stamps my passport. The immigration office will not stamp my passport until I pay. Which I cannot do as I have no money, is ignored. *(I am not prepared to tell this arsehole about the thirty thousand I have. Ten thousand I have down my shorts, the rest hidden on Mae.)* "You have to pay Immigration, Boat registration and Gusto." "So, how are you going to pay?" *(Acting innocent, I say.)* "What is Gusto?" He ignores the question. "I can go nowhere unless escorted by the army and they will only do that when it is convenient to them." "I can demand nothing." I thought this might be obvious having an armed soldier sitting next to me. Who is looking very bored and fiddling with his rifle. He escorted me over here which is fifty yards and still within the confines of the base. "Yes, yes. I understand," he said. "Tomorrow you see my boss at ten o`clock." "What, you will arrange for the army to take me?" "I don't know, but you go anyway." If this had been in the UK talking to an official, I would have exploded by now and said. "What the fuck are you talking about you moron?" But I don't, I am very conscious of the armed soldier next to me. Bolivar looks at me with a curious expression. "Are you unwell?" "You do not look well, would you like water?" I push down the rage and frustration bubbling up inside me and reply. "Yes please, that would be nice." He gets a glass of water from the cooler, drinks it himself and says. "You sit and relax for ten minutes and maybe something will get done." While shuffling papers on his desk he speaks in Spanish to his secretary. Who, along with the soldier laughs. I know it is at my expense but remain calm despite wanting to plant my fist in his smug face.

Gimenez leaves the office, and the secretary resumes her one finger typing. The soldier slouches down even further in his chair and is staring in the secretary's direction. Without looking at me he gives me a nudge and nods towards her. I look across, she is sitting at an open front desk and from where we are we can see up her skirt, she has no pants on. Every time she taps the keyboard she opens her legs then closes them. She is putting on a show. When Gimenez comes back, she stops, and the soldier sits up. Gimenez looks at me and seems surprised I am still sitting in his office. He goes over to the printer and takes out a piece of paper, studies it for a moment than hands it to the soldier. It is what the secretary has been working on. He says. "You go now." I look at the soldier, he shrugs gets up out of the chair and jerks his rifle towards the door saying, "Commando." As I got to the door Gimenez calls out. "Tomorrow if you like come back to my office, sit and relax." "What after seeing your boss?" "No, why would you go to see the director?" "Tomorrow is Friday, nothing will happen, no one does anything on a Friday". I want to say, "from what I can see no one does anything any day." But instead through gritted teeth I thank him for his help and the water *(I didn't get)* and walk out. The soldier takes me back to the camp office. While Cortez reads the paper, Gimenez had given the soldier I sit on a chair against the wall and bang my head off it. The two soldiers working at their desks stare at me, one says, "You go to, Administration Civil?" "Yes." He rolls his eyes understanding my frustration.

Cortez finishes reading and beckons me over to his desk, pointing at the paper. "You understand?" "No, I do not." He speaks to the soldier who escorted me, he asks what happened in Gimenez's office. He turns to me saying. "This is civil bullshit, no inspection yet. I am sorry, you will have to stay on the base. Years ago, I read the book, "Catch 22", which was hilarious albeit dark and disturbing. After my conversation with Gimenez, I feel as if I am now living the plot of that book and I am not finding it funny at all. For all the difficulties I have ever experienced living in England, nothing could have prepared me for what I am now suffering here in Venezuela. This morning I was naïve enough to feel positive and optimistic. Those feelings replaced by frustration and apprehension. Culture shock is the best way to describe it, in the blink of an eye my situation could change and not for the better. Living in England you have a modicum of control over your life. But here in Venezuela? Freedom is at the whim of whatever official is dealing with you. These guys do not seem tied to the concept of law and the rights of the individual as the people of the UK understand it. In the UK, there are organisations, government officials even police you can complain to, without the fear of retribution or incarceration, if you think you are being treated unfairly. There is a safety net of law and order to protect you., that is not the case here in Venezuela. I will not tell Mid this, I do not want her to worry any more than she already is.

Day 51

As usual, I wake with the dawn. By seven thirty I am ready to go to my non-existent appointment with the billy big shit harbour master. It is now eight forty-five, no one comes to take me across to the base. I think the army are tiring of babysitting me. This could be a good thing, they might try to get the civilian administration to move quicker with whatever paper work is holding things up. Maybe now with the British Consulate involved the pace might pick up.

This is probably a dream on my part. I saw the command patrol boat go out earlier so that will not be coming to get me. I am worried that if I am not seen to be trying to keep this *(fantasy)* appointment, things could get worse. They might send a fisherman to pick me up? This causes them to resent me, having to pick up the bandido from his boat, but they have no choice. It seems when the soldiers tell them *(or anybody else)* to do something; they do not have a say.

Well past eleven o'clock and I am still sitting on the boat. I think I see Silvio on the dock trying to get someone to come out to get me, but I could be mistaken? So, here I sit, morale getting lower by the minute. If I thought it would not cause me a problem, I would get the dinghy off the boat and make my way to the dock. But I don't know if I would do right or wrong? I will just have to sit this out

Its three pm when Ortega comes out to Mae on a fishing boat to take me ashore. He says, I thought you might like a shower and something to eat. Bloody right I want a shower, I have no fresh water to spare on the boat to wash with, and I am hungry. I do not have time to get my bag. When we arrive on the base, I can see something is up. There's a lot of activity. Diez, Cortez and Zapata climb into a jeep armed and wearing flak jackets they leave in a hurry. Ortega says. We are very busy today, no one can take you anywhere. Do not worry about your appointment with the director. Sarcastically I say, manana|? Ortega grins and pats me on the back. Si manana.

A young soldier, (Herez) gives me a new bar of soap. This is quite a gift, I know toiletries are scarce, even for these guys. He will not accept any money for it. He wants me to have it. After the shower and something to eat, I sit in my usual place, on the wall outside the canteen. A guy wearing sunglasses and civilian clothes approaches me. "Good afternoon," "how are you?" Warily I answer, "Fine thank you." When he takes off the sunglasses, I realise its Silvio from La Guardia. "When we saw you being arrested, we could not help you." "Torres, Hector and I watched when the gunboat took you away." "We did not know where they were taking you." "Torres made discreet enquires and found out you were here." Silvio hands me three packets of cigarettes. "Torres thought you might need these." "He worries about you and knows you like a smoke." He looks at me with concern and asks if I am all right? "Yes, I am ok." "How are you being treated?" "Not badly, and the British Consulate knows I am here." "Communication with the soldiers is sometimes difficult but other than that I am fine." Silvio promises he will help me if he can, he will try to come back on Monday. La Guardia is fifty miles away, and it's not always possible for him to travel. He is very relaxed, and I notice the soldiers do not faze him. "Do not put too much trust in these bastard's!" I say to him. "The military do not seem well liked in this part of the world?" Silvio snarls, "Anyone in the military you cannot trust."

The Consulate arrives, and I introduce him to Silvio, I explain how Silvio and the Capitan had tried to help me. Silvio writes his phone number on a piece of paper and gives it to Harry saying. "I know many people here some of whom may be of help, please call me. Harry thanks him and said he would. When Silvio leaves he says, there is more to him than meets the eye. Yes, but in a good way I think? Harry agrees. He then tells me he still has had no success in meeting the official I had seen at the immigration office. But he has an appointment to see the director of ports, the guys boss. He will try again after seeing him.

Cortez comes back at six looking tired and tense. You must go back to the boat now, tomorrow you can go to town with Vega. The only boat available to take me is a tiny unstable dinghy. Getting in it is awkward and worrying, if I fall overboard with my tablet, I will lose the ability to contact anyone.

Day 52

Eight a.m.; today I feel like a tourist, Vega has met me on the dock wearing civilian clothes. While we wait for Ernesto, *(his taxi driver friend)* to pick us up tells me his good news. He has taken a week's leave and Cortez has no objection to him being my unofficial escort during this time. Vega can pick me up in the mornings and return me to the base in the evening as I must stay on Mae at night. The day he has planned. First, we will drop my phone into a shop near the cafe run by the Frenchman which specialises in mobile repairs. While we wait, we will have coffee. Then go to his friend's small ranchero for a beer. This sounds great.

At the mobile shop the woman who looks at the phone says she will not know until Monday if someone can fix it. Vega says, "this is not a problem. We can come back then.

Let's not bother with coffee we will go to the Ranchero."
After travelling along the coast road for twenty minutes
Ernesto turns off onto a dirt track. Five minutes later he
turns off the track into a compound surrounded by trees and
overgrown vegetation. Getting out of the car I look around.
I don't know what I was expecting a Ranchero to look like?
How do I describe it?

A dilapidated double fronted building with crumbling
render. Here and there exposing the cinderblock
construction. Oversized bleached wooden doors that have
seen better days are at each end of the front of the building.
There are no windows. To the left there is a small concrete
outbuilding. To the right a collection of plastic and wooden
slatted chairs. Four of them are around a tree stump with a
square of old plywood nailed on top. The rest are scattered
about. The ground is packed earth. Even though it's only ten
a.m. there are customers. A young couple talking, and two
slightly drunk guys sitting watching them. The owner
whose name is Chuy, greets Vega and Ernesto. He is a big
guy with a beer gut wearing a stained white T-shirt, baggy
shorts, and flip-flops. Vega introduces me and Chuy gives
me a warm welcome. He calls his wife, *(Marie)* son,
(Henry) and daughter *(Sophie)* out from the house to come
and meet me. A case of beer is brought from the outbuilding
and bottles are handed out. They are cold. The outbuilding
is fitted with a refrigeration unit, the power supplied by an
extension lead from the house.

There is no bar, the case is just put onto the ground and
you just help yourself. It doesn't take long to empty it.
Everyone is having a drink with the anglaise. A car pulls up
and five more people join us. Vega's wife, his mother-in-
law, two children and the taxi driver. Lunchtime Maria
brings out a big pot of fish stew. Looking inside the pot it
doesn't look very appetising. Fish heads floating in grey
dishwater. But looks can be deceptive, it tastes all right.

Two p.m. Vega decides party time is over, it is time to go.
His family leaves and I must pay for everybody. (*At this rate
I'm thinking the money will not last long, I wasn't expecting*

to treat everyone in the bar to food and drink). While Chuy tallies the bill, I worry I won't have enough. Chuy apologises as he tells me they amount. "Nine hundred bolivar's." Even if I work out the money to the government's exchange rate, it is less than three pounds. *(About five dollars US.)* Giving him a thousand I tell him to keep the change. We leave the Ranchero and I promise Chuy and his wife I will come again.

Ernesto takes us into town and we stop outside a bar on the seafront. A fishing boat is pulled up on the beach and being unloaded of its catch. Two guys are gutting and sorting the fish which are mostly marlin. While Ernesto and I get the beers, Vega wanders over to them. He knows the guys, why wouldn't he? Calling me over, Vega points to a marlin. He wants me to buy it. One of the guys takes off its head and divides it into three large portions. Once again, I am amazed by the price, *(six hundred bolivars'.)* The look I catch the guy giving Vega makes me realise this is because Vega is military, and I don't think he likes selling his fish so cheaply. There must be at least two kg of fish in each portion. Back at the bar I give Ernesto his bag of marlin, he is delighted and thanks me. I will give some of mine to Leah for the cantina. There is no way I can keep it fresh or eat all of it. We have a few more beers and then they take me back to the base. Vega gets one of the soldiers to take me out to Mae before Ernesto drives him home.

Day 53

Two a.m.; Loud music with a heavy bass beat sounding out across the bay has woke me up. A group of people in their cars have gathered. Parked just before the base checkpoint. Peaceful protest or just a Saturday night thing? The reason I wonder is the lyrics to the songs have an underlying tone of discord and defiance. Subtle but it is there. In Venezuela KKK is a frequent chant. After the day out with Vega yesterday the reality of living conditions and poverty hit me in the face. Grenada has its problems, but the people do not live in fear of the authorities and military like here. I cannot believe the young people of this country will stand by accepting this situation. The corruption and mismanagement by the present government is robbing them of their future. Anyone with an ounce of common sense can see this, I am not a politician or academic. Who am I to judge?

Nine a.m.; Soto informs me Vega is spending the day with his family. Fair enough, it is Sunday; I wish he told me that before he left yesterday. Soto is the only one in the office. When I ask him for today's Wi-Fi code, he will not give it to me, so I can't speak to Mid. What is this guy's problem? Besides being an arsehole!

For a military outfit I find the way it works strange. Maybe Soto has a high ranking relation? A lot of the men here seem wary of him. Leaving him smirking in the office I sit on the wall. Relief from boredom and frustration comes in the form of corporal Caballero. He brings me a glass of water and sits on the wall with me. We talk as best we can through the language barrier. Even though he is off duty he stayed in the barracks last night. I think Cortez likes someone around to keep an eye on Soto when he is away.

As we talk I don't notice Soto is behind me. "Give me money Anglaise." I say no but don't turn around. "Then you give me a fishing rod from the boat." "No, you can have one when I leave here, not before." Saying something in Spanish I do not understand, I hear him giggle like a little girl. Caballero eyes widen as I feel Soto's pistol push into the base of my skull. Then he shouts "BANG! I KILL YOU ANGLAISE BASTARD!" and laughs. I see by the sickly look Caballero gives me, he doesn't find it funny at all. Because I don't react he soon stops laughing and stalks off. Caballero apologises. He out ranks Soto but says nothing., and I wonder why? I have noticed a lot of the soldiers on the base avoid him. Cortez, Garcia, and Vega seem to be the exception, but even they defer to him sometimes. Maybe he has a high ranking relation?

Caballero thinks it wise for me to go back to Mae now

and I agree. Showing no fear to Soto aggravates him, but there is no way I will let that little shit terrorise me.

On board I barbeque two marlin steaks. While I eat (*They are delicious*) I think about my problem with Soto.

Day 54

Seven thirty a.m.; So real was the dream I had, waking up to reality comes as a shock. The weather had turned in the night, and it poured with rain. Leaving the cockpit to lie on my bunk the vent above it leaked and a steady drip had soaked the mattress. At first, I thought I've pissed myself. Wouldn't have surprised me after the Soto incident.

Eight a.m.; Ortega comes to get me. On the way to the dock he asks if I am ok? "Yes, fine why?" Shaking his head, he says "Soto?" Obviously Caballero told him about yesterday. Smiling, I say "No problem, I'm ok"

When we get to the dock Vega is waiting with Ernesto. The look on his face tells me he knows about yesterday as well but says nothing. As we leave the base Vega says, we go to my casa and pick up my family. After a fifteen minute drive Ernesto turns off the road into a big housing estate through an archway bearing the same crest as on a wall at the base. Every house is the same, the small area around each one tidy. Above every front door is the crest. The people that live here are military personnel and their families. Ernesto doesn't look comfortable and waits in the car while Vega takes me inside, proudly he shows me around then hands me an empty bag to hold? The cupboard under the stairs is full of things most people can't get or afford. Taking the bag, he puts in two packets of Aruna, (The flour to make the flat bread) Sugar, Dried milk, and a packet of fresh coffee. Then hands it back. "For you my amigo, to use on the boat." Taking it, I thank him but feel guilty. These commodities are like rocking horse shit.

He puts more in another bag and when we get back in the car along with his wife, two kids and mother-in-law, he gives it to Ernesto. Who I notice accepts reluctantly. Vega is helping me, but I can't help wondering what it will cost, eventually.

At the Techno Services the woman who took the phone has bad news. It is beyond repair and costs a thousand bolivars for the privilege of finding out. She gives the phone back in bits in a plastic bag. "I might as well throw it away then? Yes, it's no good." Vega took the bag off me saying "I will get rid of it," and dropped it into his wives' bag? Then with the children she left. It seems odd? There is a rubbish bin outside the store. Why take it?

Going for coffee in the café, the owner never looks happy when Vega walks in. While sitting outside waiting. Vega gets a phone call, and rushes off with Ernesto, telling me to stay here. After five minutes I go inside to ask about my coffee, the owner seeing Vega has gone says. "These people you are with are military?" "One is." "Please be careful they are not your friend's Signor, do not trust them." Before he can say anymore Vega walks in and asks what are we talking about? The guy looks nervous, so I say "I was asking him if he has any DVDs in English for sale? Vega looks at us with suspicion." "But he hasn't," I look at the guy. "Have you?" "No Signor sorry." "We will just have the coffee then, thanks." "Will you bring it out to us?" "Si, Signor." "Well, where did you go?" I say to Vega as I steer him out the door. He wants to question the guy, but I keep talking so he comes outside and sits at the table. Ernesto is there looking miserable. While we wait his phone rings again. Flustered, he walks away to answer. The guy brings the coffee and whispers. "Thank you, remember what I said." Then, goes inside before Vega comes back. *Ernesto heard him but says nothing.* When Vega does, he is angry. He stomps inside the cafe and brings the owner out to translate. The poor guy looks like he wants the ground to open and swallow him. "Who is Silvio?" His attitude is pissing me off. "A fisherman who tried to help me." "Why?" (*I think it wise not to mention Torres).* "This person is at the commando, he has brought back your battery, also the British Consulate is on his way. Cortez wants you to return." "Ok, let us finish the coffee first."

When we get back to the base, Silvio stands up and shakes my hand while looking Vega up and down. He tells a stern looking Vega who he is and how we met, his easy going manner seems to placate him. Vega knew nothing about Silvio, I had not mentioned him. I think Vega does not like sharing me with anyone. Worried that a slice of the pie he thinks he will get, will go missing. The Latin American psyche is a bloody minefield, today I have stepped on a couple. If Vega withdraws his help now, I foresee problems. His contacts and influence would be a great loss.

Harry arrives, and I introduce him to Vega. Shaking Vega's hand and thanking him for helping his countryman, Harry asks is there anything he can do with pushing the inspection of the engine? He would appreciate any help. Vega with feathers smoothed goes to the office to speak to Cortez. Harry knows there is nothing Vega can do, but makes him feel important, Harry knows how to work with these people. Silvio says, "British diplomacy, at work." "Hello Harry nice to see you again." "Roy, I have brought back the battery fully charged." "Torres sends his regards and hopes you are well?" "I am. Thank him, please." "Ok, I must go now, take care."

Harry has a question and news. "What date did you get picked up and taken to immigration?" "Monday the twenty second of June, why?" Frowning he stares at me. "You're sure." "Yes, I wrote it in my journal." "They are saving you up for Christmas." "Pardon." "You were going to buy their Christmas presents and dinners." "I finally got to see the Port Authority Director for the area." "He did not know of any British National?" "Until I told him." "While I was in the office he telephoned the Capitan at immigration and asked why no one had sent notification of your presence? Capitan Lopez denied knowing anything about you, but he will inquire. Lopez said he would call him back, but the director told him he would wait on the phone while enquires are made. Within two minutes Lopez come back on the phone saying someone had mislaid the paperwork, he would send it. The Director said ok, but, make yourself available to see the British Consul tomorrow and have a copy ready. The Director is also sending an authorization to Gimenez for the inspection to go ahead." I ask "When?" Harry shrugs. "If nothing happens within the next two days I will ask the embassy in Caracas for permission to ask the Ambassador to intervene." "Do you have any health issues that could get this moving quicker on humanitarian grounds?" "I have arthritis and have to take meds every day, is that any good?" "Maybe." I tell him about Soto, saying, "I'm just letting you know in case something happens?" "Don't do or say anything, it would probably make matters worse?" "Are you sure?" "Yes." "Tomorrow when I have seen Lopez, if I can I will come and let you know what is happening." Not until Harry has gone does Vega reappear. Now all smiles and back to his normal jolly self. "My amigo I go home now, see you tomorrow, we go to town, yes?" "Yeah sure Vega."

Day 55

It rains during the night, and I have to sleep in the cabin. A thump on the roof disturbs me, I hear hushed voices and someone walking around. Sliding out of the bunk, I pick up the machete. There is another thump, trying not to make any noise I make my way to the steps leading up to the cockpit. Now I hear the tick over of an outboard engine. Someone is aboard Mae. Shouting, "GET OFF MY BOAT YOU BASTARDS," and rushing up the steps into the cockpit, I see a guy on the fore deck. He drops a fender and jumps into a small boat which immediately takes off. I watch the boat until it disappears into the darkness. My heart is beating fast. Not from fear but anger. Looking around, I see all but the fender the guy dropped is missing.

The time is five a.m.; There is no chance I will go back to sleep, so making tea I then sit on the aft cabin roof and stew about the loss of the fenders.

Nine a.m.; a fishing boat takes me ashore and I go to the office to report the theft. Cortez and Morales are there, embarrassed someone has robbed me right under their noses. He sends out the patrol boat to find the culprits even though I tell him not to bother.

Sitting in the cantina waiting for Vega, Ortega comes in with a message from him. He cannot come today, I will have to stay on the base. To be honest, it's a relief. Going out with him and Ernesto is draining the money. Besides paying for everything else at the end of each day, I also must pay Ernesto for the use of the car. Compared to what our days out would cost in London its pennies, but I don't know when or where I can get more until the bank transfer comes through. If Mid is right, there is still three weeks to go before the money arrives.

Sitting on the wall outside the canteen gives me a good view of any activity. A black Mercedes pulls up in front of the office. Two guys dressed in smart casual clothes, wearing wrap round sunglasses get out and look around. Cortez and Caballero rush out of the office and snap to attention in front of them and salute. They say something to Cortez, Caballero comes over to me and says, "Anglaise go to the small beach and stay out of sight." Then hurries over to the cantina, calling for soldiers to come out. Bemused, I go to the beach but position myself, so I can see what's going on. Since arriving I have seen no one on the base move so fast. Every soldier on duty forms up in front of these guys, stands to attention and salutes. One of them walks to the car and opens a rear door. A short stocky man dressed the same steps out and walks over and stands in front of the parade. The first two flanking him while he addresses the soldiers. I am too far away to hear what he is saying.

The speech goes on for fifteen minutes, then they all cheer. Leaving the soldiers standing to attention Cortez walks back to the car with him, escorted by I presume the bodyguards. Once the guy is back in the car Cortez stands to attention and salutes, holding it until they drive away. He doesn't give

131

the order for the soldiers to dismiss until the car has passed the checkpoint. Caballero comes to the beach and tells me I can go back to sit on the wall. He offers no explanation, who the guy is or why he was here? I wanted to ask why they wanted me out of the way? But let it go, maybe Vega will clue me up? For the rest of the day I sit on the wall reading. There is no word from the Consulate and since the visit from the mystery man no one has time to talk. I'm glad when I get taken back to Mae. At least I can make tea and sit on a comfortable seat.

Day 56

Nine a.m.; Harry, the consulate is on his way. While I wait I talk to Mid. Cortez gave me today's Wi-Fi code and for once it works. It's been a few days since our last chat. Mid has great news. "The bank contacted me yesterday, the transfer has gone through." "You're kidding?" "I made such a fuss at the bank the manager went out of his way and got it pushed through." "The money is in the account in Miami." "That's great babe, now I can tell Vega." "With any luck I might get it today?" "I am waiting for the Consulate to arrive, he should know more about this farce of an inspection." "When I can call you again, I will let you know what he said." Ten minutes later Harry arrives and before he says anything makes sure no one is close enough to hear our conversation. "The cat is out of the bag." "Our friend Capitan Lopez is in trouble with the Director." I must sound pleased when I say, "Really?" "Yes, you were definitely his Christmas box." "What will happen to him?" "Probably nothing." Harry says. "Seeing me was like a red rag to a bull." "He shouted at his staff and was rude to me." "Then the Director telephoned, and his attitude changed." "He became apologetic and helpful." "An order to proceed with the inspection this morning is on its way to Gimenez as we speak." "Why can't they phone him?" "He has to have a written order." "The bureaucracy here is unbelievable, everything has to be on paper in triplicate, more sometimes."

"Then it gets lost." "When I have reported a theft or an assault on a tourist, even a murder, nothing happens." "When I complain to The Chief of Police he says," "Reports of crime have to be documented and sent to me." "I have received no reports?" "If I have no reports, how can there be any crime?" "Ergo, there is no crime in my district."

Harry sighs and then carries on, "I say this, so you have an insight into what we are up against." "Forget how things work in England, none of that applies here." "Corruption is a big problem." "A lot of Civilian Officials and Military personnel are, how can I put this?" "Completely corrupt or morally pliable, Lopez falls into the first category." "Your entry stamp fee is only six dollars US, cruising fees not much more, as long as you don't stay in their waters too long." "Because of the misunderstanding once the boat is inspected, and as long as you leave within five days he has agreed no fees will be payable." "Huh, because the thieving bastard got caught out?" "Yes, but I suggest putting as much distance between here and yourself as soon as you can, before he can make more mischief for you." "You have caused him problems, and he has nothing to show for it.It's fair to say he doesn't like you."

While we talk, Vega with Soto in tow walk over. Soto looks sulky and sounds aggressive when he speaks to Vega. Harry turns away and says. "Is that the guy who threatened you?" "Yeah." "I see what you mean, I will have to watch out for him in the future, if I have any dealings with this army unit." Harry asks Vega if he can take us to Gimenez. We walk across to his office. Soto being nosy comes too. I can't go in, saying to Harry. "I will have a smoke while you talk to him if you don't mind and wait outside."

Harry looks glum when he emerges from Gimenez's office and seems reluctant to say what happened. Impatient, I say "Well?" He looks at me as if trying to guess what the reaction will be. "Bad news, the car bringing the paperwork has broken down." "The inspection will not happen today." I feel the blood rising; my head pounds. Harry looks at me, concern etched on his face. This is the final straw. After weeks of bottling up my frustrations and anger, fuck these cunts with their guns and attitude. I don't care anymore. I give in to the blind rage that engulfs me. Glaring at Soto I smack one hand with a fist imagining his face crumpling under the blow every time a word erupts from my mouth "WHAT," Smack. "IS," Smack. "WRONG," Smack. "WITH," Smack. "THIS," Smack. "FUCKING," Smack. "PLACE?" Vega and Soto back away alarmed and confused by my outburst. Harry calmly says, "Get out of bed the wrong side this morning?" Looking at him I laugh; the fury subsides as quick as it arrived. "Sorry about that." Harry smiles. "Don't worry, I feel like that every day." Vega asks Harry what is wrong with me? I interrupt and ask Harry to tell him we need to see his friend. He doesn't understand so I say, "Tell him transfer complete." Vega's eyes light up and he rushes off to organize transport. With Vega gone I ask. "What happens now?" Harry shrugs, "We wait and see what happens tomorrow." "Can you get hold of a phone?" "If all goes well with Vega's friend I will buy one today." "Good." He hands over a business card. "If the inspection is tomorrow, and you have a phone, call me yourself, if I do not hear from you I will see you the following day."

Eleven a.m.; Once again I am in the office of the guy who sits behind the huge mahogany desk. His son is with him. We wait for the woman to serve coffee and leave. Vega called him to verify the money had arrived and could we come and collect? It had, and we can, so Vega called Ernesto to bring us here. This time no one else is with us. Vega wants no witnesses to the amount of money about to change hands. The measly six hundred dollars, *(At the rate I am getting.)* converted to. One Hundred and Seventy-Seven Thousand Bolivar. Vega earns fifteen thousand a month, to him this is a small fortune. "Here is some in cash." The son hands me a bag containing the twenty-seven thousand. "The rest, we have made out a cheque payable to Vega, this is acceptable?" Vega looks at me his eyes bright. I can see the avarice in them and want to say, "No, it fucking isn't," but hold my tongue. We shake hands and leave.

Instead of going back to the base Ernesto takes us to Kate's.

She is expecting us; coffee is ready on the table, when we arrive. Vega speaks to her then she turns to me and says. "Vega knows how upset you are today and wants to explain why the cheque is made out to him and thought it best I translate, so there are no misunderstandings." "He doesn't want you to be angry." "The gentleman you dealt with would only proceed this way." Before I can ask why? She continues. "Vega wants to assure you the money is safe, tomorrow if you want he will take you to a bank and get another fifty thousand." "But he does not think it would be wise to have so much in cash." "After tomorrow anything you wish to buy he will pay for with his bank card." I drink the coffee while I think about it then say. "Tell him I trust him and realise what he says makes sense, but yes I do want the fifty thousand tomorrow." "Also, today I want to buy a phone, will this be possible?" After she translates Vega agrees and smiling excuses himself to use Kate's toilet. We will go to town today.

While he is away Kate slips a note with her telephone number on it and a message to call as soon as I can? In town getting a phone is easy. The SIM card to run it is another matter. Vega has to put it in his name and I can tell he is not happy about it. We stop at a bar and have a beer before going back to the base and I try out the phone by ringing Harry the Consulate.

Six a.m.; Back on Mae I call Kate and ask what she wants. "Ernesto and I want to warn you, Vega has big eyes, be careful." "Big eyes?" "Yes, he is a greedy man." She says when I pay Ernesto for the use of the car Vega takes half the money. Ernesto and I are not bosom buddies with Vega. He only knows me because I am Ernesto's friend. Vega uses people. Running Vega around is costing him money. This explains why he looks so miserable. I ask, "Has Vega paid you anything for translating?" "No." "Ok Kate, don't worry tell Ernesto I will make it right for the both of you tomorrow." "Tell Ernesto to pick you up in the morning before he comes to the base." "We will go to the bank together." "If Vega says anything I will say I want you along to translate in case there are any problems."

Day 57

After talking with Kate last night, I thought about Vega. All along I knew he was not helping me for my benefit. That I could accept, no one does anything for nothing. I am pissed that he is not being fair to Ernesto and Kate. I will bypass Vega and pay them myself, he is, and will, earn enough out of me as it is.

Nine a.m.; The patrol boat is leaving the dock and I can see Cortez, Gimenez, and a civilian onboard. Pulling alongside Gimenez waves a piece of paper. "This is the order for the inspection, this man is the mechanic who will carry it out." "Can we come aboard?" *Gimenez being polite makes me suspicious. What's he up to?* I lift the sole plate, so the guy has access and give him the ignition key. While he fiddles around, I call Harry and tell him what is occurring. "Who is with the mechanic from administration?" "Gimenez." "Let me speak to him." I hold out the phone to Gimenez. "The British Consul would like to speak to you." This catches him off guard, but he takes it and they speak for a few minutes then hands it back. Harry tells me to call when the inspection is over.

The mechanic finishes poking around and tries to start the engine. As I expected there is a click then nothing. Taking out a camera he takes a few pictures of the engine, writes something on a docket then informs Gimenez the engine will not start. "WELL DUH!" Gimenez then tells me "The inspection is over you are free to get a mechanic to fix it." Then giving me his most smarmy smile gives me the punch line. "Someone wrote the order for the inspection two days ago." I interrupt his flow. "No, yesterday." He looks even more smug. "No, yesterday someone sent it and unfortunately it did not get here." "We have to observe regulations." Producing the order from his pocket he points to the date and day. "You see, they signed twelve noon

Monday." "Twelve noon Saturday you must leave or face heavy penalties as you have no entry stamp on your passport." Today is Thursday, the bastard knows there is no way I can get a mechanic, let alone any parts. I cannot get the engine fixed by then.

Cortez waits while I lock up then takes me ashore. In the office he says goodbye, he is leaving the base. He has been promoted and reassigned. The new commander will arrive tomorrow. In the meantime, Vega will still escort me. As I will leave Saturday, I can stay ashore and not return to the base or Mae of a night if Vega is with me. Shaking his hand, I congratulate him on his promotion and thank him for looking after me.

While I wait for Vega, I call Harry. "They have finished?" "Yes." "Gimenez has told me I have to leave on Saturday at noon or face heavy penalties." "Why Saturday? I was led to believe you would have five days, to fix the engine?" I tell him about the order fiasco. "Ah, I thought this might happen, this smells like Capitan Lopez." "You don't think I should ignore this five day nonsense and leave when the boat is ready." "God no, leave Saturday and get away." "Without that entry stamp and the cruising fees paid you will have more trouble than before." "There will be nothing I can do to help." "Call me Saturday before you leave."

Ten thirty a.m.; Vega is ready to go to the bank, as we walk towards Ernesto's car he sees Kate sitting in the back and asks why she is here. I don't answer, but when we get in I ask her to translate. After our business at the bank I am taking them out for the day, I will pay. This quells any objections he might have. He makes a phone call then says he invited his wife to join us, but she will only stay a short time. Is that ok? Kate nudges me with her knee and rolls her eyes. When there is a free lunch on offer I expect nothing else from Vega. So, I say, "Si Vega, no problem."

The bank is a strange affair. Only Vega and I go in and we have to wait. Unbeknown to me Ernesto's wife works in the bank and has arranged everything. Taking out fifty thousand in cash is not a regular occurrence. A teller calls us over to a booth, away from prying eyes. We stand and watch while he counts the money. None of it is in bundles, all loose and of different denominations. Fifteen minutes later he finishes counting and piles the money up on the counter. Vega signs for it and I scoop it straight into my bag. I have learnt that the entrepreneurial spirit is alive and well in Venezuela. They do not allow bank staff mobile phones while they are working. The manager locks their phones away and only gives them back once they leave the bank. There has been a spate of muggings. Bank tellers would inform their accomplices of anyone leaving the bank with a sizeable amount of cash. They wait usually armed and on a motorbike for the person to come out. Being with a member of the military I will have no such problems. But Ernesto parks outside opposite the door just in case.

When we get into the car Vega's wife is sitting in the back. I ask Kate to tell Ernesto to take us somewhere nice. We go to a part of town I have not been to before. Ernesto pulls into a car park under a building that looks like a shopping mall. I'm surprised once inside by the variety of named stores. Even though the Americans have sanctions against Venezuela, I see McDonald's, Kentucky fried chicken, restaurants, and designer shops. Everyone agrees they would like a Chinese meal. I tell them to order what they want. Watching the people walking around the mall, I can see this is a place for the haves, not the have-nots.

After finishing her meal Vega's wife leaves, he walks outside to wait until her taxi arrives. This gives me a chance to talk to Ernesto and Kate on our own. Saying as I am leaving Saturday I will give them fifteen thousand bolivar each as a thank you for their help. Obviously, I cannot give it to them right now as we are in a crowded place. Ernesto asks, "do you have to return to the base tonight?" "Because if you don't you can stay at my house if you would like to." I explain, Vega is my escort, and I have to stay with him if I do not return to my boat. Kate laughs. "Vega is not one to stick to the rules I will ask him if it will be okay."

When Vega comes back she asks, and he has no objection. Then I remember I cannot leave Mae unattended overnight someone might try to board her again. Vega says he will arrange for a soldier to stay on board through the night, but I will have to pay. "How much?" "Three thousand." I take that as meaning two for the soldier and one for Vega. Smiling to myself, I think knowing him, it's probably the other way around. It's agreed but I tell him he will have to pay out of the money he's holding for me. Also, can he settle the bill for the meal? Summoning a waiter Vega with a flourish gets out his card and settles the bill.

It looks like he's treating his friends, and I get it. Besides the money Vega likes to be the generous nice guy. I can't help it, I like him and at least I'm being rinsed by a professional. Ernesto takes Kate home while Vega and I go to a bar. Even though we have a language problem and I don't completely trust him, I enjoy myself. Ernesto comes and picks us up two hours later and we go to his house. Vega only stays for ten minutes then gets Ernesto to run him home while I stay and get to know Ernesto's family.

When Ernesto returns I give him the fifteen thousand as promised and another fifteen for him to give to Kate. After dinner and a shower, I sit with the family trying to answer their questions about England but as always, the language barrier is a problem. His wife made up the bed in their spare room for me and at ten o'clock I go to sleep in a proper bed for the first time since I stayed in the room in St Martin.

Day 58

Ernesto wakes me at nine. After a shower and breakfast, he lets me connect to his Internet, so I may Skype Mid. I tell her what happened yesterday and all being well I should be back in Granada within the next three to four days. Not mentioning having serious doubts this will happen at all; I don't want to worry her. After the call I sit in Ernesto's backyard and make a list of the supplies, I will need for the trip.

Vega shows up at midday eager to go and spend more of my money, and asks Ernesto if he can stay the night? Ernesto says yes if I have no objection to Vega sharing the spare room? It's not a problem there are two single beds.

Our first port of call is the Ranchero. Chuy is pleased to see us and brings out a case of beer. The two guys from before are sitting at a table sharing one bottle. Feeling generous I ask them to join us. What the hell, this will be the last time I am here. After three hours and another case of beer Chuy's wife Maria brings out a large pot of rice and spicy chicken pieces. I eat a plate full, to soak up the beer. Then we go back to town.

At a small bar on the main street we sit outside on the patio drinking coffee along with fiery shots and watch the world go by. The barmaid is friendly and speaks English. She asks questions about life in England and says she would like to leave Venezuela but thinks this will never happen. I say it's such a shame this is such a beautiful place. "Yes, there used to be a lot of tourists, but not now." "Who wants to come to a place where there is a good chance of being ripped off, robbed, kidnapped or murdered."

Vega gets restless and wants to move on. We go to a casino and for once my luck is in, while Vega and Ernesto play the slots, I sit at a blackjack table and can't stop winning. I pull off the table with twenty thousand bolivars in chips. Cashing in is a problem, the limit is two thousand. A woman explains you must change two thousand walk away and then come back and change two more. Giving Vega and Ernesto five thousand each we alternate going up to the exchange booth. Ernesto and Vega offer the money back, I tell them to keep it.

The money is burning a hole in Vega's pocket so now we go on to a nightclub. Inside it is obvious the place is a brothel. Sitting at a table, three women come and join us. They're all good-looking and from Columbia. A waiter comes to take our drinks order, the women have fruit juice and we stick to beer. Ernesto is uncomfortable and doesn't like being here. He wants to go home, but Vega wants the woman next to him. He takes her to a private room. Ernesto goes to his car to wait. The woman next to me wants to take me to a room, but I'm not interested and have a dance with her instead. We communicate using the Google Translate on her phone. She shows me pictures of her home in Colombia where she has a young son. I ask how much she would charge to go to the room. Ten thousand bolivars is the going rate, but I can have her for seven. Happy to just sit and talk I give her four thousand bolivars for her time. Vega reappears looking worse for wear. I get the bill for the drinks and pay then we go to the car. Ernesto is drunk but still drives us back to his house. We creep in and go straight to bed, Ernesto warns us to stay quite if we wake his wife he will suffer.

Chapter 10

Adios Juan Griego

Day 59

Far off in the distance I can hear a voice calling me. "My friend, my friend." Opening my eyes sunlight is streaming through the window. Dust motes swirl around amongst the rays of light. My head is aching and feels like a lead weight as I lifted it from the pillow and turn towards the voice. Vegas's round jolly face swims into view, his jet black eyebrows raised. He is sitting on the edge of the bed next to mine. Only wearing a pair of shorts, his fat belly spilling over the waistband. Try as he might he does not look so jolly this morning. I can see his suffering like me from the previous days and nights drinking. "My friend, Ernesto has prepared breakfast we must go." Sitting up and swinging my legs over the side of the bed I feel dizzy and groan. Satisfied I am awake Vega stands up and goes to the door. I say to him, "are you okay?" Looking back, he gives me the thumbs up but has a worried look on his face. I know he is thinking of what his wife Erica will say to him when he gets home. Even the macho South American male must suffer the wife's displeasure sometimes. He gives me a sheepish smile goes out the door and closes it behind him.

I stay on the edge of the bed and give my aching head and body a few minutes to acclimatise to being vertical. I still have my jeans on from the night before, which is just as well, I do not want to bend down at this moment. It feels as if my head might fall off if I do. Voices are coming from the other side of the door, I can't put it off any longer. Pulling on a T-shirt I go to the kitchen. Ernesto greets me, "Hola Roy, coffee, Breakfast?" "Si, gracias Ernesto." He puts a Panini filled with melted cheese and ham in front of me along with a small cup of coffee, then mimes having a shower, I nod in answer. Sweeping his arm around the room he says. "Roy, my Casa your Casa." Vega looks up and speaks to Ernesto. I get the gist, Vega will have a shower first as I am still eating. I need another coffee anyway and must go outside for a cigarette. When I roll one they both laugh, shake their heads, and wag a finger at me. They believe the crazy anglaise cannot function without a cigarette stuck in his mouth. Outside in the yard I get hit by the heat. After the benefit of the air-conditioning inside the house I feel like I've walked into an oven, it is only six a.m. and the temperature is already thirty-six degrees. I know this by the gauge hanging on the wall next to the door. Ernesto's dog, a large friendly long-haired Labrador comes bounding up with a ball in its mouth wanting to play. I oblige the dog while I smoke then go back inside. Ernesto's family, who I met yesterday are up and sitting around the kitchen table. His wife, daughter-and son-in-law greet me with a friendly, "Hola." Thankfully their little granddaughter is still asleep. She is lovely but has a habit of screaming if she doesn't get whatever she wants at once. I don't think my head could handle it this morning. They do not spoil her, but the Venezuelans like the Spanish seem more tolerant of their children and indulge them more than the English. Vega has finished his shower and gone back to the bedroom. In the shower the first blast of cold water clears my woolly head. I guess with the constant heat hot water in the shower is a commodity the average household can do without. Once out of the shower I dress and say goodbye to Ernesto's family and we set off to the market. Vega wants to get there early, he thinks I will get a better deal for my provisions. It's a waste of time his friends produce is crap? All old stock and would not last the day let alone the time to sail to Granada. Now we must hang around and wait for the main market to open.

Ernesto suggests going for a coffee at the cafe near the Techno Service that tried to repair my phone. This suits me, I would like to say goodbye to the cafe owner. When we walk in the owner glances warily at Vega. I tell him I am leaving today, and he asks if they fixed my engine? He frowns and looks worried when I say no, but I have no choice they have ordered me to leave. With more bravado than I feel I say, "It does not matter I shall be all right." His wife is in the shop with their little girl and looks over with concern. She wants to say something but with Vega close by stays quiet. Ordering coffees, I then pick up a bag of sweets and hand them to the little girl. She gives me a big smile and her mother grabs my arm and kisses me on both cheeks. Going outside we sit at a table and wait in silence for the coffee. When it arrives, we drink quickly and head back to the market, where I get quality fresh vegetables and fruit. Vega as always tries to get a better deal, but I tell him not to bother, the price is fine.

Yesterday I had seen a new generator, our next job is to get it. We also pick up a full gas canister from Kate's. She refuses any payment and wishes me a safe journey. When we arrive back at the base, the new commander is there to greet me. He seems pleasant enough and insists that I accept twenty litres of petrol for my new generator. Last thing I need to do is fill my Jerry jugs with fresh water. The commander says this is not a problem. Two soldiers will take me and my gear out to Mae, bring my water containers ashore fill them and bring them back. Thanking him I then turned to Vega and Ernesto shake their hands and say my goodbyes. Vega is stony faced, and Ernesto looks upset. They know as I do this whole thing is a farce and I have a bad feeling all is not what it appears. The two soldiers helped me load everything into the patrol boat and unload at Mae. One stays aboard to help me stow gear while the other returns to shore with the Jerry jugs. Fifteen minutes later he comes back with them. Now with their job done they wish me well and leave.

It's eleven a.m. I have one hour before I leave. After making sure everything is secure and prepping Mae, I make tea and roll a cigarette. Thinking about my options the bottom line is, I have none. Without the engine repaired I am in the same position as when I arrived. As I sit smoking another cigarette, I notice to small boats with the big outboard's. Two men are in each. They do not look like fishermen and slow down as they pass by. They study Mae, then gun the engines and head out of the bay. The way they looked me over leaves me under no illusion they will wait for me when I get out into open water. It is now midday and I should be underway, but I will stall if I can. The guys in the boats did not look prepared to sit out in open water for too long, certainly not after dark.

They probably plan to jump me when I am far enough from the coast that any help could not reach me in time, and they will be long gone. This is all theory but the fact they will rob me or worse is an easy conclusion to come to. Experienced sailors who sailed the Caribbean never advertise the day or time they will leave port. The few I have spoken to who used to run in and out of Venezuela, told me they would always leave or enter at night without their running lights on. Keeping radio traffic to the minimum.

I do not have this luxury. The soldiers on the base know I must leave today at noon, along with half the fishing community. Knowing I am being watched I make myself look busy. I fuck about like this until I receive a phone call from the consulate. "Roy, you need to leave!" "The commander of the base is not happy you are still in the bay." "You will make trouble for yourself if you don't leave now." *There is no point in telling him about the two boats I believe waiting for me outside the bay, he couldn't do anything.* Instead I tell him I am having trouble with the rigging and need to run a new furling line to the genoa. Could he tell the commander I will leave within the next hour? Harry says, "I will, I wish there was more I could do?" "Be careful and good luck." Harry isn't stupid he knows I am stalling and I suspect he knows why. At twenty past three I haul up the anchor and hoist the mainsail. The bloody thing jams again halfway up the mast. The thought crosses my mind to stay and take the consequences, but I decide against it. Taking my chances with the sea and pirates seems preferable to tangling with the Venezuelan authorities again. There is no more time to mess about so, I unfurl the headsail. The wind catches it and Mae moves towards the mouth of the bay. The guys in the two boats have not come back, I hope they have got fed up with waiting and put in somewhere else along the coast. In open water wave height increases. Sitting in a small boat will be very uncomfortable in these conditions, I will keep an eye out though just in case. After being ashore so long I worry my courage will fail me, going back out to sea solo? Honestly, I am feeling wobbly, but fuck it. What will be will be? The darkness of the night is here, and I feel safer. Six miles offshore showing no lights it feels like a security blanket wrapped around Mae. There has been no sign of the two boats or anyone else. But the fact remains I still can't get to Grenada. The mobile phone still has a signal. I call Kate. It does not surprise her to hear from me and says, "Vega and Ernesto came to my house after leaving you and we discussed the situation." "Someone has set you up." "Roy it is obvious you are not going anywhere?" "Can you sail back to La Guardia before first light?" I tell her that should not be a problem the bloody wind is already taking Mae there. "That's good, we will meet you there, but you must not arrive before daybreak." "Vega has a plan." "What plan?" Kate doesn't explain and rings off. Alone and in the dark my subconscious decides this is a good time to beat me up. "This is worse than when we were becalmed." "And how do you work that one out?" "The smarmy bastard Lopez remember him?" "All the problems you have stemmed from him, and I bet this setup was to give him another chance to screw you over?" I try not to think about it and get some rest. As the night drags on sleep is spasmodic. Five minutes here ten minutes there, my subconscious always ready to stick the knife in.

Day 60

Before first light I zigzag into the bay the wind although light is for once on my side. I am three miles out when Kathy calls. "Roy get as close inshore as you can." "Vega, Ernesto, and I are coming by road to La Guardia." "Vega knows someone with a boat anchored there who will help you, His name is Julio, I will ring you back soon." Before I can ask questions, she hangs up.

Two miles from shore Mae is crawling, in the distance I can see a boat heading towards me. It is bigger than the ones I had seen the first time I was here. Kate calls again. "The boat heading towards you is Julio's, we are all on board. Julio wants you to drop your sails and secure a towrope to the bow." "Ok, no problem."

Furling the headsail from the cockpit I then get on deck to drop the jammed mainsail. Lowering it disturbs a swarm of bees and they attack. They had settled in the folds of the sail and now intent on stinging me. (*When I lived in Canada, a swarm of hornets once attacked me. My face had swollen like a football and it hurt like hell. That memory gives my feet wings*). Rushing off the deck waving my arms I throw myself into the cabin slamming the hatch shut behind me. I have not escaped there are bees in the cabin. The light from the windows that run along the side of the hull dimmed by the number of bees crawling over them. The noise outside is incredible as hundreds of angry bees try to get inside. What the hell am I going to do? The sail is not fully down. Looking in the cupboards in the galley I find a spray can of furniture polish. As I open the hatch, I hope the bees inside the cabin will follow me, they do. Shutting the hatch again while whirling the tea towel around my head, bees come at me from all sides. I keep spraying with the furniture polish and swatting them with the towel. Their attack ends as quick as it began. En mass they leave and fly towards the shore. My only explanation is a Queen must have landed on the sail and crawled into the folds. Dead and dying bees are all over the cockpit. When I finish taking down the sail, I sweep up the bodies and dump them over the side. Only stung a few times I consider myself lucky.

Julio's boat is close now. Standing on top of the aft cabin I watch its approach. Vega, Kate, and Ernesto are sitting along the starboard side. Two guys who I assume are crew on the port side. The man on the helm must be Julio. All of them must have seen me jumping around like a madman but no one comments. Julio brings the boat alongside and one of his crew jumps aboard. The crewman checks the rope I have attached to Mae's bow cleats and gives Julio the thumbs up. He then jumps back. Knowing the drill, I go to Mae's helm to keep her rudder amidships. Julio engages his engine and moves off taking up the slack. He handles his boat well and knows what he's doing. It doesn't take long to get to the anchorage he has picked out. A secluded spot opposite the mangroves two miles along the coast from the town. His crewmen help me secure Mae. Vega and Kate come over to Mae. It surprises me that Kate jumps across, she is an older lady and must have found it difficult. Vega looks on as she tells me the plan. "Julio will tow you to Concord, which is forty miles away, around the other side of the island. He will use a boat that has more powerful engines." "Once you are there, we will deal with your port of entry stamp and boat registration." "My son in Canada will pay the fees through his account."

"Along with the tow today this will cost fifty-five thousand bolivars." "You must stay here until three maybe four a m;" "Julio must move Mae before a patrol boat or worse, gunboat shows up." "No one is sure what will happen if they find you here without the right documents again." "We must legalise your status before then." "With your passport stamped, hopefully they cannot arrest you." "Ernesto and I will meet you at the immigration office in Concord tomorrow, Vega cannot be there but will be in touch." I ask, "is Vega paying Julio?" "Yes, he has arranged everything." Thanking him and Kate, I shake his hand and give her a kiss. Then they leave.

I watch the boat until it clears the head land. This wait of at least fourteen hours, will be an agony. Unlike the first Sunday I was here there are fishing boats out in the bay, but none come near me. The mobile still has a signal, I telephoned Silvio. To my relief he answers. "Hola Roy." "Julio put you near the mangroves, yes?" "How did you know?" "Kate, she rang me and told me Vega's plan." "My concern is you will arrive in Concord too late and the immigration office will be closed." "Or worse a patrol boat finds you here." "What do you think would happen if one did?" He does not answer my question, which says it all. "I know Julio, he's a good man and will get you to Concord." "Do not worry my friend, hopefully this will work out." "Have you told your lady what has happened?" "No, she has no idea and I'm not going to tell her, she would worry." He agrees this latest turn of events would drive her insane. Silvio wishes me luck and says he will see me when he can. All I can do now is wait, I have a possible fourteen nerve wracking hours to kill.

Chapter 10

Julio's Tow

Day 61

The daylight hours drag, and I feel sick. Now in the darkness the nausea is worse. While listening for the deep rumble of diesel engines fear is twisting my guts and uncertainty weaves through my mind, sowing seeds of doubt. I hear the high-pitched whine of an outboard; I know it's a fishing boat so ignore it. The small commando patrol boats do not venture out at night. The sound of big diesels would mean a gunboat is on its way. I do not know what engines Julio has on the boat he will use, but I'm sure they will not sound like those on a gunboat.

As I strain to hear every sound my nerves shred. Scanning the horizon gives me a headache. Paranoia takes hold, my subconscious bombards me with questions I cannot answer. "Will Julio keep his word?" "Is this a setup?" "What can I do if a gunboat arrives?" They swirl around inside my head with no end. All the time I am in Venezuelan territory I am still fair game. Have they let me go knowing I can go nowhere and plan to grab me while I sit in this bay. Then fit me up with drugs so they earn brownie points from their superiors. It will be easy enough for them to do this as I'm now in the first place they picked me up from. They could say I dropped something here and come back to collect it. The thought of ten years in a Venezuelan jail doesn't bear thinking about. It all sounds unbelievable but a place as corrupt and unpredictable as this, anything is possible. Then I hear the steady beat of a diesel engine. Faint and faraway, coming from the south-east, I tune out the chatter going on in my head and focus on the sound.

The first light of dawn turns the blackness of night into a grey gloom. Standing on the aft cabin I hold on to the mizzenmast, straining my eyes, peering into the distance.

Then I make out the shape of a large vessel, it is not showing any lights, it must be Julio. A gunboat of the Marine commando would have running lights on, and searchlights sweeping the water. From experience I know subtle is not their style. The engine throttles back and the boat glides up next to Mae. A crewman throws me a line and Julio and one of his crew jump across to Mae. "Buenos Dias Anglaise." He says in a hushed voice giving me a grin. Someone throws a heavy rope across to him and he gives it to the crewman who attaches it to Mae's bow. The mood is one of urgency, none of us have the desire to hang around in case the Marine commando appear.

In under ten minutes we are ready to leave, and for the third time since leaving Secret Harbour I watch a towrope snake off Mae's bow. Then feel her jerk as it snaps taut. Once it has, Julio powers up his engines, and takes our speed to six knots, then eases off the revs and speed levels off to five. With Mae's rudder centred she is not causing problems, even though when we are past the headland the sea becomes choppy. Julio turns parallel with the coastline when we are a mile from it. No patrol boat comes out to intercept us as we pass the entrance to the bay Juan Griego, and I am happier when it's behind us.

After two hours the towrope snaps. Mae stops and wallows, and I scramble up to the foredeck. As Julio turns back one of the crew pulls in the length of rope trailing behind. Ten minutes later with the rope secure again we carry on. Dark clouds are ahead of us, if the weather gets too bad Julio may cut Mae loose and I would not blame him. He cannot afford to put his crew and boat in danger.

After an hour the rope breaks again, the sea is rougher than earlier, and as Mae wallows a wave broadsides her. She heals over to port and water rushes into the cockpit, Mae quickly comes back upright, and the water pours out through her gunwales. Mae is pitching and tossing making it difficult for me to get onto the foredeck. Once there, Julio's boat comes up fast on my port quarter. He has left the helm and is standing on the bow, urging whoever is steering to get closer to Mae. I can't believe what I'm seeing. Julio is hanging on to the bow of his boat with no lifejacket, no safety line and intending to jump across to Mae. The difference is about six feet in height from the rise and fall of the boat's. I don't know if it is luck or design, but when close enough Mae goes down and Julio's boat rides up; he jumps and lands on her deck. The helmsman engages reverse and backs off fast. Julio helps me attach another towrope then waves his boat to come back. When close enough he throws the rope across to a crewman then jumps back. As soon as his feet hit the deck, the boat pulls away. Julio points ahead shouting across. "AMIGO, WE GO RAPIDO, RAPIDO." Hurrying back to the cockpit I am full of admiration and respect for Julio and his crew. All kinds of things could have gone wrong with that stunt. Taking nothing from the R.N.L.I, lifeboat crew would have been wearing all kinds of safety gear. These guys are wearing T-shirts shorts and cheeky grins. Venezuelan fishermen are not big on health and safety.

Lucky for us the weather gets no worse and the last part of our journey is uneventful. We arrive at our destination at one o'clock in the afternoon. I drop anchor near a big sailing yacht that's flying the South African flag. After securing Mae, I get onto Julio's boat and he takes me to the main jetty at Concord. Julio and his crew look happy the journey is over. I thank them for their help and shake their hands, like me relieved that we encountered no military vessels. Julio has earned fifty-five thousand bolivars' for helping me, I am sure he will give his crew a good payday. I give them a bottle of Anis to share but they refuse it and hand it back. Julio says, "you paid, and that is enough."

I watch as they head back out to sea then walked towards the building that houses immigration and the clearance office. Lopez the port Capitan watches me from the Veranda. When I reach him, he holds out his hand expecting me to take it. Greeting me like an old friend he says, "I have been expecting you." Wondering how the hell he knew I was coming I grit my teeth smile back and shake his hand,

I have to play his game. He ushers me into the building and asks for my passport and would I like a coffee? Other times I have been in this building I am ignored but now the atmosphere is different. There is a table and two chairs against a wall, he invites me to sit down and someone will bring coffee, then disappears taking my passport. A woman brings the coffee and when I try to take it outside, so I can have a cigarette a guy bars the doorway and tells me to sit back down. I study a poster on the wall opposite. At the top in big block letters is the slogan. We do not tolerate "CORRUPTION." The text underneath loosely translated says. "The government is making every effort to stamp it out, anyone caught taking bribes will face heavy penalties." Laughing to myself I wonder if the deluded soul who printed the poster knows he wasted his ink.

Ernesto and Kate walk in. "Are you okay Roy?" Kate asks. "Yes, fine but confused." Lopez appears greets them both and takes them into his office. A short time later Silvio arrives. "Where is Ernesto and Kate?" "In there with Lopez," I point to the door of his office. Silvio knocks and walks straight in, closing the door behind him. Twenty minutes later they along with Lopez come out of the office. Lopez hands me my passport, the port of entry stamp is on one of the pages.

The four of us go outside and I say. "Can one of you tell me what just happened?" Silvio explains. "Kate's son in Canada has paid the fees." "Lopez is arranging the paperwork needed for you to berth the boat in the marina Veneture." "Tomorrow he has someone who will tow your boat there." "How much does he want?" "Twenty thousand." "What to tow Mae less than a mile." "Why can't I stay anchored where I am? My reason for this is I only have thirty thousand in cash left. There cannot be a lot if anything left in Vega's bank account. While we discuss this Vega arrives.

He hands me an envelope, "my friend I have a present for you." Kate translates for him. "Inside the envelope is a one-way ticket to Trinidad leaving at four p.m. on

Wednesday which is in two days' time." "Getting a flight from here is difficult and when he heard this ticket was available, he got it for you." "There was not enough money left, so he borrowed some from his mother-in-law."

After Vega leaves I say to Silvio, "I don't want to sound ungrateful, but he's given me a problem, hasn't he?" "Yes, it may not be possible for Lopez to have the paperwork sorted in time?" Silvio goes back into the building to talk to him. When he comes out, he says "Tonight, stay on your boat, it is not wise to leave her unattended in this place." "Tomorrow before midday Lopez and his brother-in-law will move your boat to Veneture, and he will have the paperwork done." "Ernesto and I will see you there." "Now we must go, take care." Before she gets in the car Kate gives me a hug and says she will ring me tonight to check I am okay.

Watching Ernesto's car disappear down the road I savour the moment. This is the first time I have no one escorting me. I am here legally no one will arrest me. Then I realise I must get back to Mae.

Just up the dirt track that runs alongside the dock there are some ramshackle buildings. As I walk towards them I see one has tables and chairs. Outside, two attractive girls are sitting with drinks. It must be a bar, maybe someone there will have a dinghy and be willing to give me a lift out to Mae. The buildings are timber construction and they all have iron grilles over the windows, it isn't the high-end of town. It reminds me of the street Clint Eastwood rode down in the film, "High Plains Drifter."

Sat on the steps of the bar is my old friend Fernandez. The man who conned me on the exchange rate for my few dollars, the first time I came to Lopez's offices. He is drunk, nursing a bottle of beer. Walking over I sit on the steps next to him. Looking at me, eyes bleary with drink he says. "Oh, it's you, still here then?" Before I answer he launches into a rant about how I stitched him up. "You were going to call me, I had arranged with a friend of mine to exchange your money, but you never rang me?" "It was a good deal, and I would have earned a commission." I do not understand what the man is talking about? But let him waffle on. Then say. "Do you want a beer?" "If so, stop fucking moaning and I will get you one." He jerks his thumb at the barred window behind us. "You can get one from there." Shouting, "Dos Cervasa Por Favor," I bang on the grill. A flap I had not noticed opens, a voice says. "Cien bolivar." I put the cash into a hand that appears, seconds later someone passes out two bottles of beer. "Very trusting around here?" Fernandez says most of the businesses here operate like this because of the high risk of being robbed. "Not a place for tourists especially after dark." *If he is trying to worry me, it isn't working.* "Suppose not, but from what I've seen of this place everywhere is like it."

We drink in silence for a while, then he says. "Your friends, the soldiers, they accused me of being a thief and I should not have ripped you off when I changed your dollars." "They gave me a hard time." Acting surprised I say. "This is news, but still you dodgy bastard, you did rip me off." Fernandez looks at me hurt and offended. His clothes are not far off being rags, he looks old and worn out and I can't help feeling sorry for him. "How much would you have got in commission?" "Five hundred bolivars for every hundred US dollars." It was hardly a fortune, I give him the money to get another beer. While he does this, I peel off five one hundred notes from the role I have and give them to him when he comes.

Stuffing the notes into a pocket he then points to a small yacht. "That used to be my boat., Someone offered a good price, so I sold it." "With the money I set up as a tour guide, but things went wrong and it never worked out." "Did you live on the boat?" "Yes, but now I have a room next to that restaurant." Fernandez points to a wooden building on the edge of the dock. The only indication it is a restaurant is a poster pinned to a tree outside. While trying to convince me he has plans, I read between the lines, he'd drank the money for the boat, and now catches any crumbs that fall from Lopez's table. A black Mercedes convertible glides towards us, pulling up just down the road. A smart dressed young guy gets out and walks over to the two girls sitting opposite. Seeing the guy Fernandez becomes twitchy. "Do you know him?" Before he can answer the guy walks over and speaks to him. From the tone he asks Fernandez who I am? When Fernandez answers he sounds timid and subservient. The guy speaking English with an American accent says. "Hi, Fernandez tells me you are from London?" "Yes, that's right." "Why are you here?" Not wanting to go into details I say. "Waiting to get my boat moved to Veneture so I can repair her." He laughs and shakes his head. "Good luck with that, I hope you're not in a hurry?" Then goes back to his car, whistles to the girls, they get in and he drives off. "Friend of yours Fernandez."

Sneering he says, "No, that man is not my friend." Then he goes quiet. Time is marching on and I need to get out to Mae. "Do you know anyone who will take me out to my boat?" Seeing the chance of getting more money Fernandez walks to the end of the dock. After shouting himself hoarse he attracts a guy's attention on the big yacht flying the South African flag. Coming over to the dock in a small dinghy he ties it off but doesn't get out of it. Fernandez gabbles away in Spanish. I hope asking him if he will take me out to my boat, not setting me up. Even though I have spent the last couple of hours with him I don't trust him as far as I can spit. The guys looking up at me not saying a word. He looks like a down and out, with his raggedy clothes. Despite this, he looks fit and healthy. When Fernandez stops talking he says. "Hi, my name is Michael, you are English and need a ride?" "Yes." "No problem, you ready to go now?" I give Fernandez another two hundred bolivar's then get in the dinghy. Michael looks relieved to be leaving the shore. He says hanging around the dock at night is not something you want to be doing. Fernandez disappears before we are halfway to Mae, he must think the same. When we get to Mae, I ask him if he would like to come aboard for a coffee. "You have coffee?" "Yes, but only powdered milk to go with it if you take it white." "Yes please, I haven't had coffee for weeks, but I have to go back to my boat first I will be back in ten minutes." While he goes to his boat, I make the coffee and stick a machete under a cushion in the cockpit. It seems paranoid, but I have learnt not to trust anyone. I want a weapon close by just in case all is not what it seems.

When Michael comes back, I give him coffee and offer a cigarette. Taking a deep drag a smile spreads across his face and when he sips the coffee it turns into a grin. "How long have you been here?" He asks. "To bloody long, I can't wait to get out of here." "I know how you feel Venezuela can seem like a madhouse." "But it is a beautiful country and most of its people are honest and friendly."

He points across the bay. "You see that." Buildings outline the skyline, below them the odd street lamp glows. Car headlights flash between the trees lining the road beyond the beach, which looks like a ribbon of darkness separating the land from the sea. I follow his line of sight as he tracks his finger along the beach. "There!" "Did you see that?" Bright light grows out of the darkness, someone has lit a fire. A figure passes in front of the flames, obscuring the light. "What's that about then?" "Over there is a shanty town, Illegal squatters." "Most are honest, scratching a living as best they can." "But they are also desperate, they have very little money or food." "If an opportunity arises to rob an unattended yacht, they will take it." "I advise you to keep a light on, it usually puts them off if they think someone is onboard." "Usually?" I say raising my eyebrows. "Well most of the time, but if they sense weakness, they may chance it." "There have been violent incidents." "When I first stayed on the yacht I had to make it plain, I would not stand for any shit and defend the boat." "they understood the message." I slid out the machete from under the cushion and held it up. With what I hoped was a hard look on my face. "Yup, I think that would make your position quite clear." We both laugh. "You will hear boats through the night, you may get the odd one come close and check you out." "Our boats are close together, so, will leave you alone."

Michael speaks English with the guttural twang of a South African, I ask him from what part is he from? "Cape Town, but I haven't been back for years." "Not by choice I have been here since nineteen ninety-nine." He looks like he has a story to tell and trying to decide if he can trust me with it. "I was young and foolish, I got caught trying to smuggle drugs and sentenced to seven years." "There was no remission." Tensing he looks at me, I suspect waiting for a sign of disapproval or moral outrage. "Was you using a boat?" "No, I was assured I would get through customs at the airport." "I know, I can see it on your face, how could I have been so stupid, I have asked myself that question many times." "So why are you still here?" "The Venezuelans took away my passport, I was using a false British one." "When I was released, I thought they would deport me, but they didn't." "Without a passport I cannot leave." "There is nothing for me in South Africa anyway, my family disowned me when I was arrested." "Twice I have applied for Venezuelan citizenship and been refused." "I am on my third attempt." "Why didn't your government help you?" "I only saw the consul twice in all the time I was in prison." "He could do nothing, only inform the family where I was, and they did not want to know." "Anyway, my life is here now, it's not so bad." He looks across to his boat. "The yacht is not mine, I look after it for a small fee and it gives me somewhere to live." "If I get my papers this time my situation will improve." Before he goes back to the yacht, we have another coffee and I give him a pack of cigarettes, he tells me to shout out if I get any problems.

Day 62

Last night I slept on and off for about eight hours. Only waking twice when I heard an outboard engine. Sitting on the aft cabin roof eating breakfast in the sunshine could be idyllic if it wasn't for the circumstances.

Eleven a.m.; Lopez arrives with his brother-in-law to

tow Mae to Veneture. When he comes aboard Mae, his eyes are everywhere. Hiding the resentment, I feel of this arsehole on the boat is difficult, but I hide it. Half an hour later with Mae berthed, Lopez goes back to Pampatar with his brother-in-law and promises to have the documents ready by four. He says I can collect when coming to pay the twenty thousand. I had made him aware he wasn't getting the money until I had them.

At one pm Silvio and Ernesto come to the marine. Silvio says it would be in my interest to strip Mae of any equipment I can take back to the UK with me, and what I can't he will store at his house. Ernesto says if I want I can stay at his place until I fly out. Agreeing to both these suggestions Silvio and I start to strip gear from the boat and Ernesto loads it into his car. It takes until dusk and a few trips, the dinghies and outboard we leave till last. But once these are off her there is not a lot more we can remove. Silvio warns me even though Mae is in this marine she is at risk until I can come back for her. That's a chance I will have to take. Silvio tells me he will look after Mae the best he can. Then it's back to Ernesto's, I must speak to Mid. She knows nothing about any of this and will have to arrange a ticket for a flight from Trinidad. I only hope she can?

Chapter 11

Airports, Planes and Delays

Day 63

Mid has booked me on a flight leaving Trinidad for Gatwick at eight thirty pm tonight. The flight from here to Trinidad only takes twenty minutes. I should make my connection with plenty of time to spare? Kate has come to the airport to act as interpreter in case of any problems. We haven't been here ten minutes before I get the first one.

Ernesto sits in the waiting area while Kate and I join a long queue to check in. A woman is walking along the line with a clipboard asking for names and checking them off on the flight manifest. Kate gives her my name, and as she looks down the sheet, a frown appears on her face. Mine is not on it. Thank God Kate is with me, I do not understand a word the woman says. After what seems like an argument they resolve the problem.

It takes twenty minutes to get to the front of the queue, where a man sitting behind the counter has a raft of forms. Kate does the talking, and he hands me two. I must fill them out then join another queue to hand them in. With this done a soldier puts my bags on a conveyor belt that runs through an x-ray machine. Two others take them off the other end and they make me unpack them. They check everything questioning what the radio and Gps I took from Mae are for? Satisfied with my answers they let me repack. Then put the bags on a trolley and tell me to wait for the departure lounge to open. Kate and Ernesto must leave. Kate says. "You should be ok now, when you get back to England please call and let us know when you get home. Promising I will I shake Ernesto's hand and give her a hug, thanking them for everything.

Three forty five the departure lounge doors open. I think they are cutting it fine. The plane is scheduled to take off at four. Four thirty I am still sitting in the lounge. There has been no announcement about the delay. At the flight desk, I ask, "What is the problem?" The guy gives me a blank stare? "If I don't get to Trinidad soon, I will miss my connection." This gets a response. "Do not worry, we will call ahead and ask for your flight to wait." Now I stare at him, the arseholes face doesn't crack. Sarcasm is all I have left. "Seriously, you expect me to swallow that? Under my breath I add, you fucking moron. Angrily he waves me away, telling me to sit down and wait. I'm sure he heard and understood.

The guy I sit next to says. "You will get nowhere; the plane will leave when it leaves." "It is always like this; these people do not give a shit." "You should not have booked your connection for the same day." A soldier struts into the lounge and calls out six names, followed by a short speech which I don't understand. Mine is one of them. The guy next to me says "Is your name on his list?" "Yes." "You have to go with him." The other five people on the list have stood up, the soldier is looking around for the sixth. When I stand up, he leads us out of the lounge to another room. He calls out one name then after telling the rest of us to sit he takes the person through a door leading out of the building. No one speaks, but everyone looks nervous. There is a fifteen minute interval between each time he takes someone out, I am left until last.

When my turn comes I walk through the door and see my bags on a table. Two soldiers stand behind it and two more stand either side. The soldier on the left has a small dog. The one on the right is holding an assault rifle. In most countries I would have nothing to worry about, but here in Venezuela? My stomach rolls over and fear tickles the back of my neck. As I walk towards the table I think, *someone has planted drugs in my bags and a big show of finding them will now take place.* I am told to open the bags and take everything out. While I do this the soldiers laugh and joke amongst themselves. The two behind the table then go through my things. Taking apart anything that can be. Then they encouraged the dog to jump on the table and sniff around. Terrified it will find something I hold my breath while trying to look unconcerned. The dog finds nothing, and I am told to put everything back. Then taken to the plane where only three of the others called out are waiting to get on board. Twenty minutes later we are in the air. Not until I feel the wheels touch the runway at Piarco International do I feel relief of sorts. Because of the delay I have another problem, I have missed my flight to Gatwick.

Stepping off the plane, I follow the crowd and collect my bags from the reclaim hall. The atmosphere here is so different. The underlying sense of fear and frustrations of the people is missing. People look happy. There are no armed soldiers, *(Not in view anyway)* stalking around intimidating people with their stares. Only when I join the queue for passport control does unease creep up on me. How is the Trinidad immigration official going to react to the fact I have no ongoing flight or return ticket?

As I get closer to the head of the queue, I study the woman checking passports, trying to gauge her attitude. What is Trinidad's stance on illegal immigrants? Will they put me on a plane back to where I've come from? When my turn comes, I hand her my passport and explain the circumstances as to why I have missed my flight. Frowning, but never interrupting the flow of words gushing from my mouth she listens. When I finish she makes a phone call, then hands back my passport and tells me to wait. Someone is coming to deal with me. A young woman wearing the uniform of the Trinidad Border Authority approaches me. Smiling she introduces herself.

Her name is Elise. Could she see my passport? While she flicks through the pages, she asks me to repeat what I told her colleague. Feeling like a wimp, before I finish I add, "Please don't send me back to Venezuela." She says, "This happens from time to time with passengers from Venezuela, as you are a British citizen I do not think it likely you will have to go back." "Wait here and I will see what I can do?" While I wait the nagging words of doubt chew on my brain. "Will they?" "Won't They?" When Elise comes back, she brings a trolley for my bags and tells me to follow her. "Stay with me, technically you should not pass through control, but as you are with me, it will be ok." We go through the doors that open into the arrivals hall. "Your onward flight to the UK was with Caribbean Airlines, we are going to their reception desk." Elise tells the woman on reception what happened and gives her my passport. After bringing up my details on the computer she says. "Because you missed the flight, you must pay for another ticket.". "That is not possible I can't afford it; besides it is not my fault I missed the flight." She must hear the desperation in my voice. "I will talk to my supervisor, please wait." Ten minutes later she's back. "He has to talk to admin, but no one will be available until tomorrow." Elise asks, "What do you think they will say?" The woman shrugs. Elise takes me over to a seat. "Wait here I need to speak to my boss, would you like a coffee?" "Yes please." "I will bring one when I get back."

She marches off, and I think how different the officials are here, compared to the ones in Venezuela. Or have I hit lucky? When Elise comes, she has brought coffee and good news. Her boss has been on to someone at the airline. We go back to the woman on the desk, she is just putting down the phone. "My supervisor has confirmed you do not have to pay for another flight and they will issue a ticket at check-in, providing you pay the administration cost." "How much is that, and when is the flight?" "One hundred and fifty dollars." My face crumples. "What, American dollars?" The woman laughs. "No, Trinidad dollars, and the earliest flight I can get you on is tomorrow at seven pm."

Elise takes me to an ATM. When I put in my card, in my mind I beg the machine to give me some money. I have no idea if there are any funds in my account? Unlike the machines in Venezuela this one does not want passport numbers. Not knowing the exchange rate or prices of anything here I ask Elise what she thinks would be a reasonable amount to get. Bearing in mind I have a long wait. "Five hundred, a thousand? Without waiting for her answer, I tap one thousand. The machine, clicks and whirrs then spits out my card, PLEASE WAIT FOR YOUR MONEY appears on the screen.

On the way back to the desk Elise says, "You must stay within the confines of the airport and the concourse outside until you leave." "What, like Tom Hanks in The Terminal?" She grins, "Yes." "No problem, I will, and thank you so much." Ok, I will leave you now. Go and pay the admin charge and safe journey, After I pay I find a left luggage place and dump my bags, I don't want to keep lugging them around. Then I order a meal in one of the restaurants. While I wait for the food I connect with the airport Wi-Fi, so I can call Mid. I must let her know I missed the original flight, and the new time I will leave Trinidad and arrive in Gatwick. She says, "Ok I will be there to pick you up." "That's a long time you have to wait in the airport." "Don't fall asleep and miss it." I promise her there is no way I will miss the flight.

Day 64

Six pm; Exhausted as I am I have not slept in the last twenty hours. The adrenalin coursing through my body and excitement of going home stopped me from sleeping. Now after boarding the plane, buckling up the seat belt and settling in my seat I don't stop it from taking me.

Day 65

The flight passed with me dead to the world; a stewardess has to wake me up when the plane lands. Coming out of the arrivals hall I see Mid amongst the crowd gathered to meet their friends and families. Mids feelings show on her face, the disbelief that the skinny hollowed eyed wreck walking towards her is the man she left in Grenada. When we get home, my brother is waiting to see me. After looking me over he says, "Well bruv that trip done you the world of good. You look seventy years old and by the colour of your hair I have to ask." "Has a dog been pissing on your head the last six months?"

The End

roycleeter.com